# Process Development Division



Held at the 2023 AIChE Annual Meeting

Orlando, Florida, USA
5-10 November 2023

ISBN: 978-1-7138-9305-9

**Printed from e-media with permission by:**

Curran Associates, Inc.
57 Morehouse Lane
Red Hook, NY 12571

**Some format issues inherent in the e-media version may also appear in this print version.**



Printed with permission by Curran Associates, Inc. (2024)

For permission requests, please contact AIChE
at the address below.

AIChE
120 Wall Street, FL 23
New York, NY 10005-4020

Phone: (800) 242-4363
Fax: (203) 775-5177

www.aiche.org

**Additional copies of this publication are available from:**

Curran Associates, Inc.
57 Morehouse Lane
Red Hook, NY 12571 USA
Phone: 845-758-0400
Fax: 845-758-2633
Email: curran@proceedings.com
Web: www.proceedings.com

# TABLE OF CONTENTS

**LABORATORY AND PILOT PLANT SAFETY**



**POSTER SESSION: PROCESS DEVELOPMENT**

## PROCESS DEVELOPMENT: DESIGN, RISK REDUCTION, IMPLEMENTATION AND OPERATIONS



## PROCESS INTENSIFICATION - NOVEL REACTORS

## PROCESS INTENSIFICATION – NOVEL INTEGRATION CONCEPTS



## PROCESS INTENSIFICATION – NOVEL TECHNOLOGIES FOR CARBON CAPTURE AND CARBON RECYCLING



## DECARBONIZATION FOR ALKENE PRODUCTION

**DECARBONIZATION FOR METHANOL PRODUCTION**



**PROCESS RESEARCH FOR IMPROVED THROUGHPUT AND EFFICIENCY, AND REDUCED COST**

## MANUFACTURING TECHNOLOGY IMPROVEMENTS FOR CHEMICAL/ENERGY INDUSTRIES



## MATERIALS AND PROCESSES FOR WATER PURIFICATION AND DESALINATION

**DECARBONIZATION FOR AMMONIA PRODUCTION**



**DECARBONIZATION FOR HYDROGEN/CHEMICALS PRODUCTION AND CO2 VALORIZATION**



**CHALLENGES AND BEST PRACTICES IN TECHNOLOGY COMMERCIALIZATION II**

# 76a Utilizing Digital Transformation to Accelerate Technology Commercialization

*Anam Ahmed*

Expanding competitive landscape has placed a sense of urgency in specialty chemical companies to continuously innovate and expand their product portfolio. This requires multiple time intensive processes such as R&D for new products, debottlenecking studies, project implementation, trial production runs, product qualifications which is then followed by full-scale commercial production.

In order to overcome these challenges, many companies are leveraging digitals tools, to expedite the process, achieve production agility and reduce lead times. However, there are several caveats. Proper integration and deployment of digital tools to reach their full potential require consideration of several factors such as: 1) Adaptability of the tools, 2) User-friendly interfaces, 3) Cost in upkeep and training, 4) long-term applicability of the tool, 5) Pilot studies for model, 6) Adaptation phase and much more. These pose significant challenges when scaling the technology to achieve the economic benefits.

This talk will go over the best strategies used to leverage certain digital tools to accelerate innovation, realize resilient and autonomous production with specific examples and scenarios. It will also address the challenges faced when deploying and integrating digital tools to increase innovation, capacity, and provide strategies to avoid pitfalls.

## 76b Modular Process Development (versus Designing Modular Process Equipment)

*Robert Nunley*

Modular design is a term often used to describe modular equipment mounted on a skid or framework so that it can be constructed remotely and moved to a job site. Recently, this has been extended into the concept of "numbering up", in which modular designs are repeated to reconstruct the same process so that it can be built and operated in multiple locations or in multiple trains.

This talk will instead focus on the concept of "modular process development". An ongoing demonstration project is taking the concept of operating modular, skid mounted unit operations on a pilot plant scale so that they can be quickly rearranged to validate process designs without having to build new pilot plants from scratch or to make major modifications to existing facilities. This concept is intended to speed the process of moving from bench chemistry into piloting and through validation. The skids are large enough to provide scale-up data, and if the production requirement is not large, the same skids could be used to quickly begin manufacturing specialized chemicals. Challenges encountered in this effort include the need to design equipment flexible enough to handle a variety of process conditions without understanding exactly what chemistry and process conditions it could see.

The concept for this project was derived from the COVID pandemic, with the focus being on process development when critical raw materials must be developed quickly and moved to production in a short timeframe. The project is funded by a NIST grant to the AIChE RAPID institute. AVN Corporation is leading the skid design, construction, and demonstration along with accompanying laboratory work for demonstration chemistries. Cheryl Teich, of Teich Process Development, has worked to document best practices in process development, and Nima Yazdanpanah of Procegence has provided guidance in the identification of chemicals of interest.

## 76c Characterization of Non-Traditional Reactors for Scale up of Highly Branched Polymers

*Jennifer Larimer*

Turbulent immiscible liquid-liquid systems produce functionalized highly branched polymers via hydrolysis and condensation reactions. Due to simultaneous and interdependent physics, including coalescence, dispersion, heat and mass transfer, and competitive consecutive chemical reaction, time dependent mixing characteristics impact the resulting material produced.

Well characterized mixing equipment such as stirred tank reactors have well established rules for designing and scaling mixing dependent processes. Specific guidelines exist for both geometric and operational parameters such as load volumes, tank and impeller diameters, impeller types, number and location, and reactant addition strategies. Pump loop systems, however, have not been explicitly studied in the literature; therefore, scaling guidance for these systems is limited. Parallels can be drawn between well characterized stirred tank systems and pump loop operation, however, there are more independent design and operational variables in pump loop systems that increase the level of difficulty in process scaling over that of the stirred tank design.

Additionally, highly branched polymers can grow in three dimensional space as a function of time. Reaction conditions as a function of time influence how the structure of these species build[1]. Different mixing conditions throughout different zones of a pump loop system can shift the instantaneous growth of a molecule between different reaction mechanisms. Therefore understanding what aspects of pump loop system design have the largest influence in how molecules build is essential to successfully scaling a process from lab to commercial scale.

### REFERENCES

1. Brinker, C.J., "Hydrolysis and Condensation of Silicates: Effects on Structure", Journal of Non-Crystalline Solids, V100, 1988, Pg 31-50, North Holland, Amsterdam. Elsevier Science Publishers B.V.

## 76d Multiphase Reaction/Separation Processes; Technologies and Apparatus Design

*Georg Rudelstorfer, Rafaela Greil, Dominik Wickenhauser, Maximilian Neubauer, Matthäus Siebenhofer, Susanne Lux, Annika Grafschafter*

Industrial scale stirred liquid-liquid extraction columns are designed to provide perfect mixing of two immiscible liquid phases. Industrial scale reactors can't be scaled up linearly from lab experiments. They often suffer from insufficient mixing caused by scale up errors. In recent time, high performance CFD simulations does help to understand the mixing in large size reaction vessels and multiphase contactors. In liquid-liquid extraction, high effort was spent on providing adequate rules for mixing from batch to industrial scale size. Scale up algorithms are available and proven. This knowledge has been applied in the design of perfectly mixed "multiphase reaction vessels" based on the design principle of extraction columns. Mixing units in extraction columns are divided in individual compartments. This compartments can be seen as multiple continuously stirred tanks, arranged vertically to obtain space and equipment savings for a cascade of reaction vessels. In this study multiphase Taylor Vortex flow (gas, liquid, solid) with a continuous liquid carrier phase in a Taylor-Couette Disc Contactor (TCDC) column setup was investigated. Taylor Vortex flow is induced and stabilized via discs mounted on a rotating shaft. The applicability of stable multiphase flow was demonstrated in three different applications. Firstly, the characterization of continuous gas-liquid flow was performed. The dispersed gas phase holdup and residence time distribution were evaluated and modelled. Practical application of gas-liquid flow in the Taylor-Couette Disc Contactor was tested by continuous neutralization of sodium hydroxide with gaseous $CO_2$. 0.1 molar sodium hydroxide solution was neutralized with a mixture of 30 vol % $CO_2$ and 70 vol% $N_2$. The process parameters rotational speed, temperature and gas flow rate were varied. To achieve a fixed pH of 9 at the column outlet the feed flowrate of sodium hydroxide was adjusted. The reaction was modelled based on the two film theory in order to provide a simple scale up algorithm. The outcome of this first study was used for implementing precipitation of lithium carbonate from aqueous lithium hydroxide solution with carbon dioxide in a second project phase. The solid lithium carbonate particles were collected in a sedimentation zone at the bottom of the column and continuously discharged with a peristaltic pump. Target of this study was the investigation of the effect of operation conditions on the particle size distribution. The results confirm applicability of continuous gas-liquid-solid flow in a TCDC column. In a next step liquid-liquid-solid operation by heterogeneously catalyzed esterification of acetic acid with methanol combined with in-situ solvent extraction of the reaction product methyl acetate was successfully performed and modeled. The hydraulic parameters were investigated in terms of residence time distribution, catalyst holdup and dispersed phase holdup [1]. The outcome of these investigations demonstrates applicability of the Taylor-Couette Disc Contactor for flexible continuous reactive and extractive operations. Based on this experience the continuous synthesis of 2-Methoxyhydroquinone (MHQ), which can be used as redox-active molecule for organic redox flow batteries [2], by oxidation of vanillin with hydrogen peroxide was applied. This synthesis reaction was performed with sodiumpercarbonate as hydrogen peroxide source. Conversion of vanillin leads to phase separation after reaction with the product acting as solvent phase. After optimization the process was operated in lab scale (50 mm diameter TCDC) for several hundred hours with a capacity of 0.6 kg/h MHQ. This paper presents the application of the TCDC as industrial scale reaction vessel. The concept is demonstrated in two environmentally friendly processes. The continuous synthesis of an organic electroactive molecule for redox flow batteries and the continuous recovery of lithium from battery scrap using CO2.

[1] G. Rudelstorfer, M. Neubauer, M. Siebenhofer, S. Lux, A. Grafschafter, Chemie Ing. Tech. 2022. DOI: 10.1002/CITE.202100184.

[2] W. Schlemmer, P. Nothdurft, A. Petzold, G. Riess, P. Frühwirt, M. Schmallegger, G. Gescheidt-Demner, R. Fischer, S. A. Freunberger, W. Kern, et al., Angew. Chemie Int. Ed. 2020, 59 (51), 22943–22946. DOI:10.1002/anie.202008253.

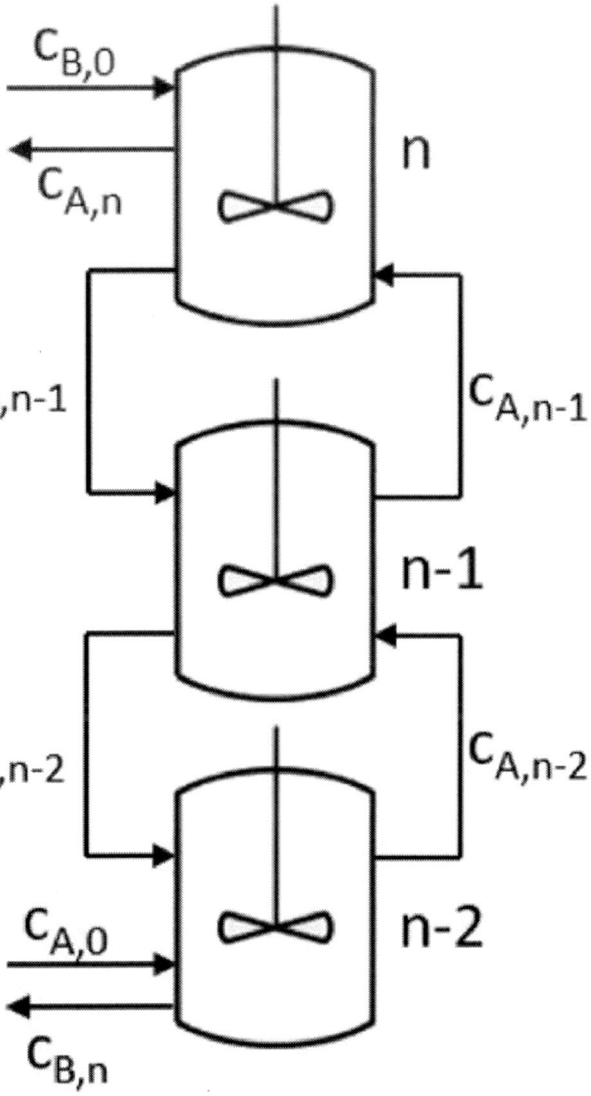

## 9a Development and Optimization of a Continuous API Manufacturing Process

*Cameron Armstrong, Katharina Grohowalski, Kakasaheb Y. Nandiwale, David Bernhardson, Pablo Jose Cabrera Ventura, Sergei Tcyrulnikov, Alexander Hesketh, David Limburg, Steven M. Guinness*

Continuous manufacturing is often leveraged in the production of active pharmaceutical ingredients (API) and intermediates to enable process intensification, cycle-time reduction, or hazard mitigation. The Flexible API Supply Technology (FAST) group is an engineering focused team that develops continuous manufacturing technology platforms in parallel with assessing the viability of and carrying out continuous processes for portfolio projects in Pfizer's pipeline alongside our process chemistry partners. We present the development of a continuous process along with the application of various automated platforms for reaction optimization in flow applying the use of dynamic/transient flow strategies for ramping residence times and reagent equivalents.

## 9b Composition-Property Relationships of BP-1 Lunar Regolith Binders for in-Situ Resource Utilization As Geopolymers and *Planetopolymers*

*Thaddeus Egnaczyk, Quent Hartt, Jennifer Mills, Norman J. Wagner*

In situ resource utilization (ISRU) is required for long-term human habitation on the Moon and Mars, with the need to construct environmental protection and critical infrastructure foremost[1]. Analogous to terrestrial construction materials formed from aluminosilicates, i.e., geopolymers, binders formed from lunar aluminosilicate regolith (planetopolymers) are necessary to create construction materials for lunar landing pads and habitats. The work presented here builds on a previous lateral study across multiple regolith simulants to focus on Black Point-1 (BP-1) lunar regolith simulant and study the effect of binder composition, i.e., compositional variations in solids weight percent of lunar regolith simulant, activating solution silicon and sodium content, and activating solution water content on seven-day compressive strength, strain to fail, and secant modulus, resulting in varying binder properties. Decreasing water content had a positive effect on compressive strength and secant modulus of planetopolymers formed at ambient-earth conditions, reaching maximum compressive strength at 75 wt% BP-1. Above a compositional limit of $Si/H_2O$=0.6, differences in chemical coordination cause variations in compressive strength with the elemental composition of the binder. At high solids in formulation, nonmonotonic behavior in compressive strength was observed with varying silicon and aluminum content, emphasizing the need for a mechanistic understanding of the relationship between composition and structure for optimization of material properties. This study provides a robust set of data for a widely available lunar regolith simulant that can guide further development of processing strategies for lunar ISRU.

[1] Comparison of lunar and Martian regolith simulant-based geopolymer cements formed by alkali-activation for in-situ resource utilization: J. N. Mills, M. Katzarova and N. J. Wagner Advances in Space Research 2022 Vol. 69 Issue 1 Pages 761-777

**0.3 (60 $H_2O$)** 55 wt% BP-1 **0.5 (60 $H_2O$)** 55 wt% BP-1 **0.5 (45 $H_2O$)** 55 wt% BP-1

**Fig. 9: Fracture behavior of BP-1geopolymer cubes formed from the (A) 0.3 Si/Σ (60 H2O), (B) 0.5 Si/Σ (60 H2O), and (C) 0.5 Si/Σ (45 H2O) activating solution at a constant composition of 55 wt% BP-1.**

## 9c Sustainable Manufacturing and Life Cycle Assessment of MOF-Polymer Composites By Reactive Extrusion

*Yufeng Quan, Harold Escobar Hernandez, Qingsheng Wang*

Over the past decades, significant research has shown that manufacturing MOF-polymer composites is a promising strategy for improving its practical adaptability in storage, separation, and catalysis applications since unshaped MOF powders are not appropriate for deployment in engineering processes. However, there are some gaps between laboratory-synthesized and commercial-deployed MOF-polymer composites: most MOFs are synthesized in lab in milligram/gram scale under batch reactions and are then used to manufacture composites in a discontinuous process; it requires a large volume of hazardous solvents during the solvothermal synthesis, which posts more environmental issues and increases material costs. In this work, we propose a continuously one-step, rapid, and solvent-free approach to manufacture MOF-polymer composite directly from solid reagents (polymer matrix, metal irons, and ligands) via extrusion that can reach a few kilograms per hour scale compared with conventional method in grams/batch/day. For the first time, this environmental-friendly approach has been demonstrated in a case study of ZIF-67 using Process 11 parallel twin-screw extruder. Further LCA studies focus on identifying and studying the sustainability of MOF technology, specifically the potential industrial production at large scales. By comparing the production of 1 kg ZIF-67, reactive extrusion exhibits several orders of magnitude (~3-5) lower than solvothermal reaction in terms of environmental and health impacts.

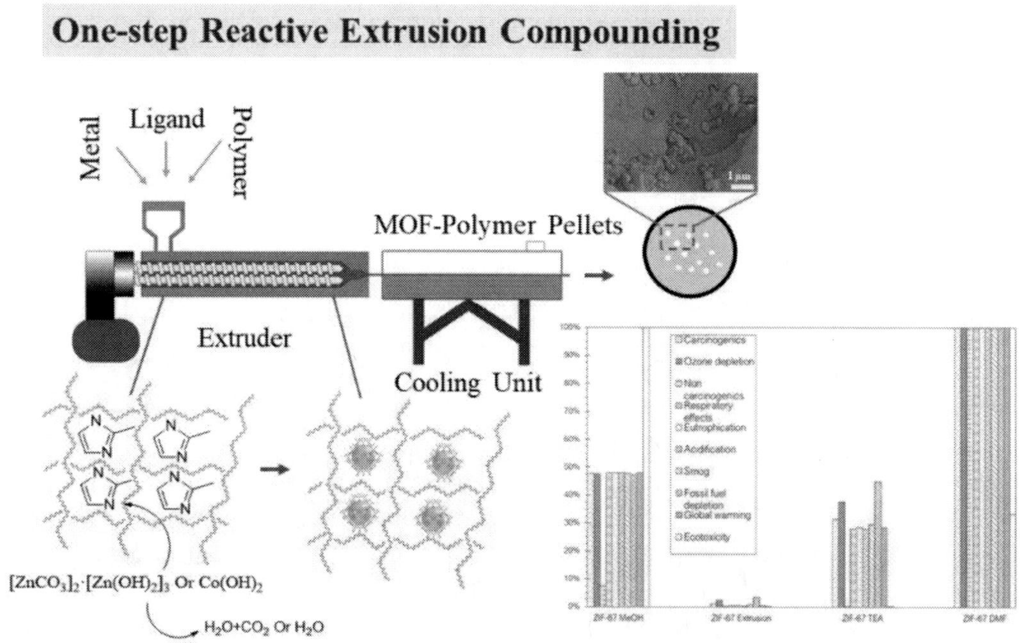

## 9d Demonstration of a Sustainable Pathway for Producing Fully Bio-Based Polyethylene Terephthalate (bio-PET)

*Parikshit Sarda, Sridhar Viamajala, Joseph Lawrence*

We report the demonstration of a commercially scalable, environmentally sustainable, and economically viable pathway for producing *fully bio-based* polyethylene terephthalate (PET). Commercially available bio-PET contains only ~30% bio-content – the ethylene glycol is made from bioethanol, but terephthalic acid continues to be derived from fossil sources. In contrast, our approach uses low-quality fatty acids and triglycerides (e.g. waste cooking oils and brown grease) as feedstocks. Our process uses a novel, patented, and easily scalable vapor phase thermo-catalytic conversion of the waste oils to obtain high aromatic yields. Subsequently, the p-xylene is separated and converted to high purity terephthalic acid and polymerized with bio-ethylene glycol to obtain a 100% biobased PET product. Other methods to obtain bio-p-xylene rely on lignocellulosic biomass. Direct conversion of lignocellulose (e.g. via pyrolysis) results in low p-xylene yields and a complex mixture of xylene-like products that makes p-xylene purification difficult and potentially prohibitively expensive. As an alternative, production of p-xylene from lignocellulosic sugars or lignin have been explored; however, these approaches are predicated on commercially viable lignocellulosic biorefineries that do not yet exist. In contrast, waste oils are currently available, collected, and could be easily be incorporated into our process for bio-PET.

Our innovation for thermo-catalytic conversion of waste oil involves delivery of the feed into a hot reactor via an atomizer to allow the oils to vaporize rapidly and facilitate catalytic reactions in the vapor phase. As a result, our pyrolysis reactor generates products with high yields, high selectivity, and low coke formation on the catalyst. At the optimum reactor temperature of 500°C and a weight hourly space velocity of 6 $h^{-1}$, we have measured organic liquid product yields of 63% (relative to feed mass) with an aromatics content of 48%. Further, we have shown *in-situ catalyst regeneration and reuse for 10 reaction runs without measurable impact on product yield*; the duration of each run was 30 min. The product spectrum is similar to the mixture obtained from commercial petrochemical production and allows for easy p-xylene separation. We will present data on upstream conversion, downstream purification, and polymer synthesis to demonstrate a fully viable pathway for bio-PET by our process (see Figure below). Property comparisons of bio-PET and petro-PET will also be presented.

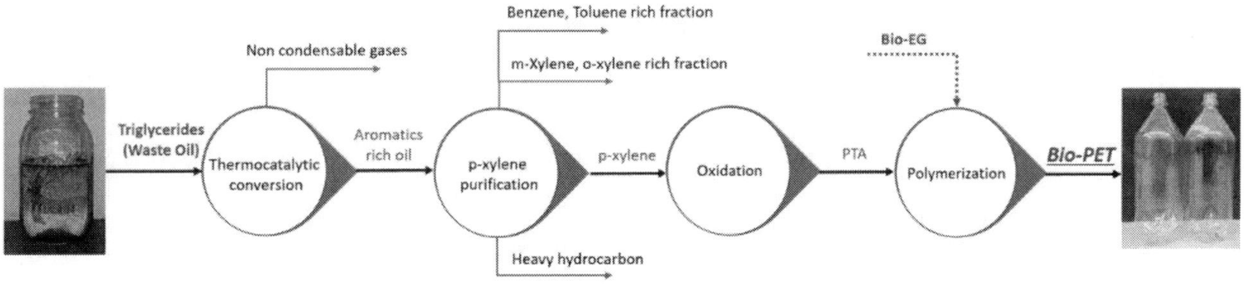

## 9e Sustainable One-Pot Production of Diformylxylose from Agricultural Biomass: Process Development, Scaling-up, and Life-Cycle Assessment

*Anastasiia Komarova, Zezhong Li, Jeremy Luterbacher*

The manufacturing and disposal of chemicals are among the primary contributors to carbon emissions and environmental pollution. Despite widespread efforts, more than 99 % of the most common chemicals are still not produced sustainably. At the same time, the top 20 global chemical manufacturers have publicly committed to reach near-zero emissions or carbon neutrality by 2050. Addressing these challenges requires a collective effort to develop and adopt more environmentally friendly and safe chemical processes and products with minimal carbon impact. Biomass plays an important role in this transition since biomass is the largest source of renewable carbon on Earth. Agricultural waste, particularly corn cobs, is one of the most promising biomass feedstocks since it is available in huge quantities, does not compete with food crops, and is usually burnt or buried in the fields. With over 1 billion tonnes of corn produced globally each year, corn cobs represent an advantageous source of renewable carbon for the production of carbohydrate-based platform chemicals.

Diformylxylose (DFX), a new xylose-based platform chemical, demonstrated potential as a versatile and sustainable alternative in various applications where it can serve as a starting material for the production of xylitol in a tripled yield, a precursor for furfural production in biphasic system, as well as a polar aprotic etheric solvent with comparable performance to toxic and environmentally harmful solvents (e.g. NMP, DMAc, DMF).

In this work, we demonstrated the one-pot production of DFX from D-xylose and corn cobs in high yields, while adhering to the principles of green chemistry. The process was successfully developed and scaled up from lab- to multi-kg-scale. We performed continuous process simulations and techno-economic analysis, providing insight into the viability of DFX production at different scales and scenarios. We also conducted a preliminary life-cycle assessment and revealed that DFX offers a very low global warming potential as well as minimal environmental impact. Finally, we showed that DFX is inherently biodegradable which facilitates its disposal. Overall, this study provides a practical example of how waste biomass can be sustainably transformed into valuable platform chemical with low cost and high efficacy.

# 9f Selection of Greener Solvents Using Computer-Aided Molecular Design (CAMD) to Extract Octacosanol from Filter Press Mud of Sugarcane

*Nistala Venkata Subrahmanyam, Sharad Bhartiya, Urmila Diwekar*

## Introduction

Generally, sugar industries generate large amounts of solid wastes like bagasse, rind, press mud, and bagasse fly ash. Filter press mud is obtained during the filtration of the cane juice performed after the clarification process. Sugarcane mills in Brazil, India, and China account for 75% of the global supply of filter press mud [1]. In addition to esters, aldehydes, and carboxylic acids, acids, press mud consists of high-value nutraceuticals like policosanols, which are aliphatic alcohols consisting of an even number of carbon atoms (C26-C32), of which 1-octacosanol is the most abundant, and are known to have appreciable health benefits. Octacosanol has been associated with health benefits such as inflammatory diseases, cholesterol-lowering effect, *anti-Parkinsonism*, etc., [2]. For these reasons, perhaps, pure 1-octacosanol is a high-economic-value nutraceutical, and considerable interest exists in separating octacosanol from the mixture of compounds in filter press mud. This work assumes that the octacosanol purification will start with a solvent extraction process. We focus on arriving at a rational choice for the solvent for extraction of octacosanol from filter press mud.

Initially, this paper begins with a review of the compounds reported to be present in sugarcane solid wastes. A study of the analysis of various solvents using Hansen's solubility theory is presented next. The main contribution of the paper consists of exploring the space of solvents, characterized by the various functional groups, via CAMD and optimization in order to obtain those solvents that exhibit the highest solubility. An optimization approach to CAMD involves a combinatorial optimization method, as the resulting problem involves the combinatorial explosion of alternatives. This poses a challenge to optimization techniques. The possibility of obtaining local minima or maxima is very high in such problems [3]. Some of the biomimick optimization techniques, like ant colony optimization, show a higher probability of obtaining global solutions. We are using an efficient ant colony optimization technique to solve the CAMD problem of solvent selection.

## Octacosanol in Sugarcane

Octacosanol is one of the compounds present in policosanols, which are a mixture of linear α-carbon fatty alcohols. It is hydrophobic in nature ('fatty' alcohol), which implies it is insoluble in water [4]. Attard *et al.* [5] reported that rind is rich in saturated fatty alcohols (i.e., policosanols), saturated fatty aldehydes, and a little in wax esters and fatty acids. Table 1 shows some of the compounds rich in rind.

In this study, we assume that the policosanols present in rind are transferred to the juice during crushing and subsequently manifest as filter press mud during clarification. Thus, the compounds reported in Table 1 may be assumed to be the primary constituents of filter press mud, of which octacosanol must be recovered and purified.

Recent studies on the extraction of policosanols make use of solvent extraction with a Soxhlet apparatus and supercritical fluid extraction [5]. Supercritical fluid extraction is a high-cost process due to which its application is limited [6]. Subcritical liquefied extraction, an advancement of supercritical fluid extraction, costs lower energy and is a lower-pressure process. A recent study with this technique with dimethyl ether as solvent yielded 249.21 ± 13.9 mg of C28 per 100 g of the crude extract [6].

Since the compounds present in policosanols are homologous (i.e., differ only by $-CH_2-$), their properties are similar to each other. Therefore, we would at least try to recover the maximum amount of octacosanol from the filter press mud using a solvent extraction process. Thus, the ideal solvent should have high solubility for octacosanol to yield high concentrations of the extract and differential affinity for octacosanol so that the extract also exhibits high purity. If high purity is attained via the solvent, then the burden on the downstream purification of octacosanol will be lower.

**Solvent Selection: Hansen Solubility Theory**

Conventionally, the solubility of solvents is determined by experimental tests made in laboratories. Although this method is reliable, it is based on trial and error, and a better solvent may be missed out. Also, this method does not explicitly account for constraints like environmental and health safety. Another rational approach is based on the contribution of individual functional groups and exploits the availability of a database of the properties of these individual groups [3]. Hansen solubility parameter model is a method where the solubility parameter (δ), which was coined by Hildebrand and Scott [7], serves to determine the solubility of a given solute in a given solvent [3]. Solubility δ is the square root of the cohesive energy density, $(E/V)^{1/2}$, where E denotes the energy of vaporization and V denotes the molar volume of pure solvent [8].

The square of the solubility parameter ($\delta^2$) is the sum of squares of three other terms which denote dispersion ($\delta_d^2$), polarity ($\delta_p^2$), and hydrogen bonding ($\delta_h^2$) between groups of molecules. This denotes the sum of individual cohesion energy densities equals total cohesion energy density.

Based on the solubility parameters of the two compounds, a new parameter distance (Ra) was introduced by Skaarup to study the interactions between the two compounds. The three-dimensional solubility parameters are calculated based on the types of groups present in the compound. Unless the distance between solute and solvent is less than that of solute and raffinate, solute molecules are immiscible (or simply insoluble) in the selected solvent. Liquids having close values of solubility parameters are soluble with each other, which in turn results in high distribution coefficient. High values of distribution coefficient get rid of large size of extracting equipment, and the requirement of recycling solvent lowers too [3]. Manuel Diaz de los Ríos*et al.* [9] came up with this technique, Hansen solubility theory, to obtain a fraction of sugarcane wax using ethanol as solvent. They concluded that ethanol was the best solvent at high temperatures because the temperature effect and other thermodynamic factors, namely dissolution and mixing, revealed that affinity to attract increased. In addition, a mixture of ethanol and water are taken in 5% v/v and 10% v/v, respectively, was capable of extracting aldehydes and alcohols with increasing temperature. This important study showed the path to another application of Hansen's solubility theory and the dependence of temperature and other thermodynamic factors in this theory. However, the study was limited to extracting a fraction of sugarcane wax using ethanol would be possible only under certain conditions. And this was not carried forward to verify other possible solvents that might work even under normal conditions.

The Hansen solubility theory proposes solvents made up of different groups like $-CH_3$, $-CH_2-$, -OH, -COO, -COOH, >C=C<, -O-, etc. that can make up to large combinations of solvents. Computer-aided Molecular Design (CAMD) is the reverse use of the group contribution method, like the Hansen solubility theory, to generate molecules with desirable properties. Selecting the optimal solvent optimization provides the platform where combinatorial optimization algorithms can be used to achieve the requirement.

**Efficient Ant-Colony Optimization**

A combinatorial optimization problem can be solved using an ant colony optimization (ACO) algorithm or any other metaheuristic approaches like simulated annealing and genetic algorithm [10]. However, the area of research to extend this algorithm to continuous and mixed variable non-linear problems has been extensively widened. Such research led to the development of a new variant of ACO called the efficient ant-colony optimization (EACO) method. [10]

ACO algorithm was developed by bio-mimicking the behavior of real ants that are in search of food. An ant releases a chemical called pheromone on its path and returns to the initial point. The same would help the other ants to track down the path toward the food source. There would be many paths, and as time flies concentration of pheromones in less frequently used paths decreases. This would automatically lead to shorter paths, and finally, the shortest path will be discovered.

Now, based on this idea and the inclusion of Hammersley sequence sampling (HSS), the EACO algorithm was developed. This method can solve convex and concave optimization problems and is highly efficient compared to ACO alone. In ACO, the solution archive is first initialized and generates new solutions iteratively until it finds the best one. This solution archive produces a probability distribution of the promising solutions across the search space. The performance of this algorithm greatly depends on initializing the solution archive. Because of HSS's in-built multi-dimensional uniformity, incorporating it in the ACO algorithm helps to avoid clustering of the initial solution archive and movement of decision variables. [10]

## Conclusion

Solvent selection involves finding an optimal solvent that maximizes properties like distribution coefficient and solvent selectivity. The solvent can be high boiling or low boiling. To make the solvent environmentally benign, we are including environmental constraints in the problem. In this paper, we are using an optimization approach based on efficient ant colony optimization to solve the real-world problem of finding optimal solvents to extract octacosanol from filter press mud of sugarcane. We plan to verify the optimal solvent selection results obtained by CAMD theory using laboratory scale experiments.

**Table 1** Compounds present in rind reported from gas chromatography [5]

| Compound | Rind (μg / g of dry plant) |
|---|---|
| Tetracosanol | 4.5 ± 0.4 |
| Hexacosanol | 129.1 ± 14.5 |
| Octacosanol | 1902.8 ± 242.3 |
| Hexacosanal | 137.5 ± 19.8 |
| Octacosanal | 3001.5 ± 694.8 |
| Triacontanal | 825.6 ± 315.9 |

**References**


1. FAO, 2011. FAOSTAT. Food and Agriculture Organization of the United Nations
2. S. Sen Gupta, M. Ghosh.Octacosanol educes physico-chemical attributes, release and bioavailability as modified nanocrystals, European Journal of Pharmaceutics and Biopharmaceutics (2017), doi: http://dx.doi.org/10.1016/j.ejpb.2017.06.020.
3. Ki-Joo Kim, Urmila M Diwekar, and Kevin G Joback. Greener Solvent Selection Under Uncertainty. American Chemical Symposium Series 819 – Clean Solvents: Alternative Media for Chemical Reactions and Processing. (2022) 224-237.
4. Hu, B. , Li, H. , Wang, Q. , Tan, Y. , Chen, R. , Li, J. , Ban, W. and Liang, L. (2018) Production and Utilization of L-Arabinose in China. World Journal of Engineering and Technology, 6, 24-36. doi: 10.4236/wjet.2018.63B004.
5. Thomas M. Attard, C.Rob McElroy, Camila A. Rezende, Igor Polikarpov, James H. Clark, Andrew J. Hunt. Sugarcane waste as a valuable source of lipophilic molecules. Industrial Crops and Products 76 (2015) 95-103. doi: http://dx.doi.org/10.1016/j.indcrop.2015.05.077
6. Kamchonemenukool, S.; Ho, C.-T.; Boonnoun, P.; Li, S.; Pan, M.-H.; Klangpetch, W.; Weerawatanakorn, M. High Levels of Policosanols and Phytosterols from Sugar Mill Waste by Subcritical Liquefied Dimethyl Ether. Foods 2022, 11, 2937. https://doi.org/10.3390/ foods11192937.
7. Hildebrand, J.; Scott, R.L. Regular Solutions, Prentice Hall; Englewood Cliffs, NJ, 1962
8. Charles M Hansen. Hansen Solubility Parameters – A User's Handbook 2nd Ed.2007
9. Manuel Díaz de los Ríos, Eduardo Hernández Ramos, Víctor González Canavaciolo, Roxana Vicente Murillo, Katherine Pérez Carrión, and Lourdes Zumalacarregui de

Cárdenas.Obtaining a Fraction of Sugarcane Wax Rich in Policosanol by Using Ethanol as Solvent: Results Interpretation through Hansen's Solubility Theory.ACS Omega 2022 7 (31), 27324-27333doi: 10.1021/acsomega.2c02314

10. Diwekar UM, Gebreslassie BH (2016) Efficient Ant Colony Optimization (EACO) Algorithm for Deterministic Optimization. Int J Swarm Intel Evol Comput 5: 131. doi: 10.4172/2090-4908.1000131

## 54a Resources for Teaching Process Safety That Are Managed or Have Been Developed By the Center for Chemical Process Safety

*Bruce Vaughen, Dr.Anil Gokhale*

Process safety focuses on reducing the risks associated with both the toxic, reactive, flammable, and explosive properties of materials, and hazardous operating conditions such as high pressures or temperatures. AIChE, through its Undergraduate Process Safety Learning Initiative (UPSLI), has developed three major process safety teaching programs designed to help 'pave the way for those currently tasked with developing and improving process safety courses.' The three UPSLI programs are (1) online curricula using the Safety and Chemical Engineering Education (SAChE) modules, (2) in-person workshops to build faculty process safety competence, and (3) in-person undergraduate process safety bootcamps to build student competence.

This presentation will introduce each of the UPSLI initiatives and discuss how the Center for Chemical Process Safety (CCPS), through its network of industrial process safety experts, has helped develop these resources for academia. The CCPS resources include process safety-related tools that can be used by instructors and students (as well as professionals) when performing process safety studies.

The first area is the SAChE Certificate Program. This program is a cooperative effort between the CCPS and engineering schools that provides process safety teaching materials for undergraduate or recently graduated engineers. After thirty years of collaboration, the 30+ SAChE modules have evolved into interactive learning modules, with a dynamic curriculum that makes incorporating process safety education into standard engineering curricula easy and engaging.

The second UPSLI area is the Faculty Workshops. CCPS, through AIChE, partners with industry leaders to provide faculty with Process Safety Workshops at sponsoring companies. These workshops are hosted at an industrial plant, showing instructors how process safety is put into industrial practice and demonstrating how important process safety is to the design and operation of the plant.

The third UPSLI area is the Process Safety Student Bootcamps. These boot camps are hosted at the school, introducing engineering students to methods that companies use to manage process safety. In addition, these boot camps demonstrate how important process safety is to both process and equipment design. Again, like the faculty workshops, the boot camps are supported by resources through the CCPS.

This presentation will conclude by introducing additional process safety tools that can be used in academia to help teach process safety. These tools, supported or developed by CCPS, include the Chemical Reactivity Worksheet (CRW), the Chemical Hazards Engineering Fundamentals (CHEF) Calculation Aid, and the publication of the textbook *Process Safety for Engineers – An Introduction, Second Edition* through Wiley in 2022. A few years ago, CCPS created its CCPSf Process Safety Fundamentals Certification Program. The CCPSf specifies 24 of the SAChE modules that must be passed to earn the certification. In addition, the CCPS has created several seven-minute videos describing process safety designs for specific equipment or introducing a process safety concept.

## 54b Increasing Chemical Engineering Students' Awareness of New out-of-Textbook Safety Topics

*Jack Shultz, Jared Rhodes, Joontaek Park*

Many chemical engineering programs have a safety course based on chemical process safety topics using the Crowl and Louvar textbook. However, there are new out-of-textbook safety topics which chemical engineers may encounter and should be aware of due to rapidly changing technologies. For example, cyberattacks on chemical plants are a newly growing issue. Therefore, cybersecurity is an important new area of focus for chemical plants requiring added attention. Chemical engineering students may also have careers in the energy or semiconductor industry, which have different safety aspects from typical wet chemical or petroleum plants. The safety course of the Chemical and Biomedical Department of the University of Missouri provides a chance for students to be exposed to these new topics through case study report and presentation. Students in small groups are assigned one of the incidents related to these topics to write a report and present on in class. They are asked to analyze the case which critical decision led to the incident based on the decision matrix [CEP Magazine (2020) July p47]. Through this practice, students gain more awareness of broad range of safety issues. We present a list of the incident cases used in class and discuss the efficacy of these projects for the education of students.

## 54c Bringing the Real World into the Classroom: A Risk-Based Assessment of the Train Derailment in East Palestine, Ohio

*Obioma Uche*

Process safety in the Chemical Engineering capstone design series at the Rochester Institute of Technology has typically been limited to HAZOP (Hazards and Operability) study of the designed units. Although this has been deemed sufficient to meet ABET requirements, we were interested in expanding our focus to include a risk-based analysis of contemporary incidents which would help students link concepts discussed in the classroom to those in the real world. To that end, our student teams were tasked with performing a risk assessment of the release of toxic gases caused by derailment of the Norfolk Southern freight train in East Palestine, Ohio. Their assessment of the incident proceeded along the following fronts: (1) determination of the risk level associated with the incident; and (2) preparation of a risk management plan for transportation of hazardous chemicals by railroad. Over the course of the group activity, the students learned about risk matrix evaluation and the risk management program administered by the Environmental Protection Agency. The structure of the assignment and associated reading materials as well as student performance and feedback from survey questions will be discussed in the presentation.

## 54d Understanding Engineering Students' Approaches to Process Safety Judgements through a Digital Process Safety Game

*Cayla Ritz, Brittany Butler, Jeffrey Stransky, Cheryl A. Bodnar, Emily Dringenberg, Elif Miskioglu*

Chemical engineers have a key responsibility in mitigating process safety incidents. When they occur, these incidents can have severe consequences (e.g., health and safety impacts on personnel or community) and are often the end result of a series of process safety judgements. To better prepare engineering students for the responsibility of preventing process safety incidents, chemical engineering programs have implemented a variety of educational strategies such as requiring process safety courses as part of the curriculum or including process safety concepts in engineering design courses. In most cases this instruction has been limited to focusing on safety and hazard mitigation and does not include how to approach making process safety judgements which are also a critical component to incident prevention.

Game-based learning has emerged as a way to simulate judgement-making scenarios for students in contexts they cannot authentically explore in a classroom. In this work, we use Contents Under Pressure (CUP), a digital process safety game, where the player assumes the role of a chemical plant manager and makes a series of process safety judgements over the course of 15 in-game days. Ongoing research with this tool has been focused on exploring the differences between engineering students beliefs and simulated behavior when making judgements. Research has indicated that if our beliefs and behaviors are not in alignment that poor judgement can surface as a result. Our data collection methodology provides us with the opportunity to collect students' beliefs (pre-CUP interview), simulated behaviors (CUP), and reflection (post-CUP survey) to directly compare student perceptions of how they approach process safety judgements in an easy-to-view format.

We observed that in some cases, students' beliefs and simulated behaviors were not in alignment, indicating that students' beliefs may not accurately predict their future behaviors. Often, students drew on past experiences to articulate their beliefs, suggesting that they were engaging in self-reflection of their past behaviors to inform future beliefs. However, many students did not have relevant experience in a chemical plant setting, so not all experiences were completely transferable. We believe that CUP provides a simulated experience with making process safety judgements that will allow students to practice making process safety judgements and aid with forming their beliefs. Thus, we want to share our approach for comparing students' beliefs and behaviors as a classroom-based learning tool.

We have developed two potential approaches for the integration of our approach into chemical engineering classrooms. The first approach focuses on individual reflection where the student will record their beliefs, capture their associated decisions within CUP, and complete a post-game reflection. For maximum learning benefit, students should then discuss their experiences with peers within the class after these activities. The second approach involves an instructor-led discussion on how students believe they will approach process safety judgements, followed by playing through a portion of the game as a whole class. Students are then encouraged to share what they observed as differences between beliefs that they held and their approaches to making judgements as part of the game. We believe that providing instructors with a variety of implementation options will allow them to customize the content to best fit their particular students' needs.

## 54e Teaching Process Safety with Simulink Simulations of Runaway Reactions

*Akhil Gopinath, Spandhana Gonuguntla, Aycan Hacioglu*

A typical process safety course curriculum covers topics such as toxicology, industrial hygiene, chemical reactivity hazards, fires and explosions, relief systems, hazard and risk analysis, and regulations. Typical calculations include critical pressure and flow calculations for compressible fluids and safe distance calculations from overpressure development.

To make process safety concepts more interactive and engaging, we developed a Simulink simulation of a runaway reaction. Simulink is a widely used tool in the industry for designing, and testing safety-critical control systems before automatically generating code and deploying it into the physical platform. In Simulink, systems can be represented as graphical and functional blocks, and dynamic systems can be simulated. Many chemical engineering instructors use Simulink in their process control courses.

This simulation allows students to change various process parameters in the runaway reaction. Students observe the reactor temperature and conversion in the reactor over time to identify the runaway reaction. Later, we add a temperature controller to the simulation to demonstrate the importance of process control and process simulation for process safety. This course material could be used as a part of process safety, reaction kinetics, chemical engineering laboratory, or process control courses.

# Development of an Elective Process Safety Course at the University of Michigan

Kenneth First, Center for Chemical Process Safety
Dr Henry Wang, University of Michigan
Dr Christopher Barr, University of Michigan

## *Abstract*

*The chemical industries have an excellent track record for safety. However, these operations pose risk and catastrophic incidents have occurred. For example, in December of 2007, a runaway reaction led to an explosion at the T2 Laboratories in Jacksonville, Florida. Following this incident, the U.S. Chemical Safety and Hazard Investigation Board (CSB) recommended improvement of the undergraduate Chemical Engineering curriculum in the United States. In response, the Accreditation Board for Engineering and Technology (ABET) updated the requirements for Chemical Engineering programs noting that "the curriculum must include the engineering application of basic sciences to the design, analysis, and control of chemical, physical, and/or biological processes, including the hazards associated with these processes."*

*The University of Michigan has updated its Chemical Engineering curriculum to include process safety related homework modules for each core course and incorporate process safety concepts into the senior Design Course such as Inherently Safer Design. In addition, an elective Process Safety course (ChE 407) was developed beginning in 2015. This industrial course applies the fundamentals of Process Safety to identify potential hazards, anticipate process upsets or equipment failures, predict the possible severity of the associated incidents and, propose means to prevent, control, or mitigate process risks. As this course is intended to prepare students for a real-world environment, various speakers from industry are invited to present first-hand knowledge of industrial process safety, as well.*

## Incorporation of Process Safety into Core Chemical Engineering Courses

As with many universities, the Chemical Engineering curriculum has little room to add required courses. To ensure that all Chemical Engineering students have some exposure to Process Safety, homework modules have been added to each of the core courses. A web page has been developed to navigate the various Process Safety Modules and Reference Information. A student may access the various modules, tutorials, and other reference information from the SAFECHE website initially developed by Dr. Scott Fogler and now managed by Dr. Chris Barr.

*Figure 1 – SAFECHE Web Page*

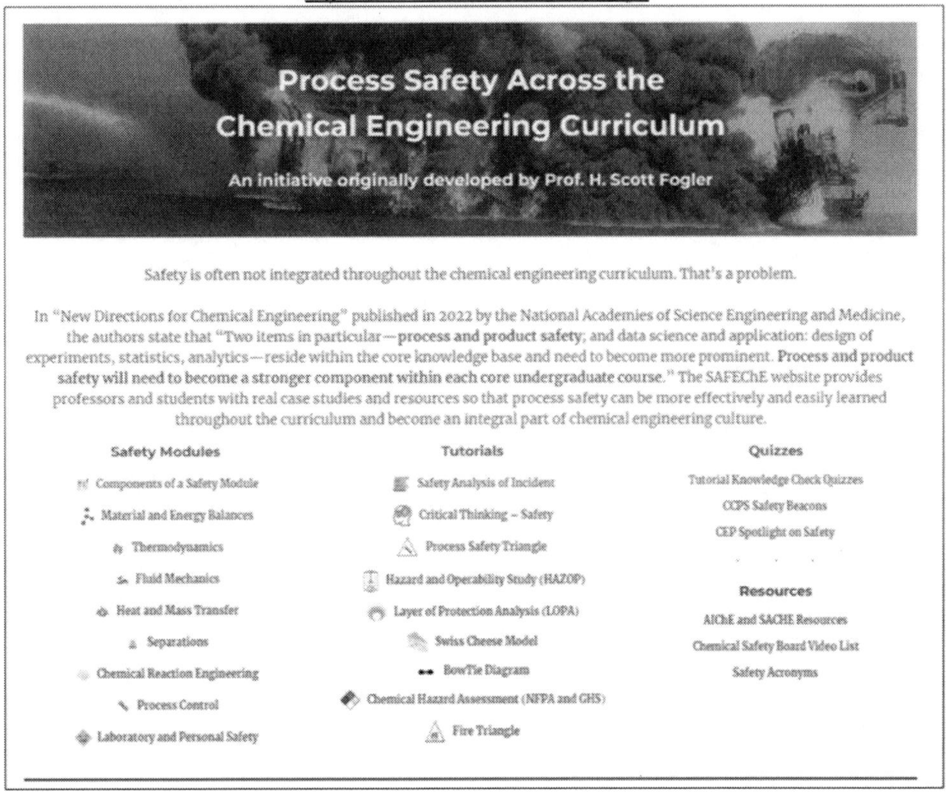


Each SAFECHE module is based on a Process Safety Incident that has been reviewed by the US Chemical Safety Board. Students watch a video of the incident (and possibly read portions of the CSB report), address several questions, and perform simple evaluations.

An example module for the Material and Energy Balance course is US Chemical Safety Board investigation of the 2009 Caribbean Petroleum Company (CAPECO) Explosion.

*Figure 2 – SAFECHE CAPECO Explosion Module*

## Problem Statement

**Safety Module 1:** *Explosion at Caribbean Petroleum Company (CAPECO), October 23, 2009, in Bayamón, Puerto Rico. Developed in collaboration with Kara Steshetz, University of Michigan and Professor Bryan Goldsmith, University of Michigan.*

This incident occurred while gasoline was being unloaded from a tanker vessel. The main storage tank was full, so the flow was diverted to one of two smaller storage tanks. The use of a faulty manual level monitoring system on a storage tank resulted in the overflow, ignition, and explosion of 200,000 gallons of gasoline, culminating in a fire which continued for 2 days. Over 300 homes and businesses were damaged and nearby soil, waterways, and wetlands were contaminated.

## Safety Analysis

It is important that chemical engineers understand what the accident was, why it happened, and how it could have been prevented to ensure similar accidents may be prevented. Performing a safety analysis to the accident will help achieve this goal. In order to become familiar with a strategy for accident awareness and prevention, view the Chemical Safety Board video on the explosion at the Caribbean Petroleum Company and fill out the safety analysis table below. See definitions on the last page. If necessary, view pages 9-13 and 25-28 of the CSB CAPECO incident report.

| Criteria | Responses |
|---|---|
| Activity | |
| Hazard | |
| Incident | |
| Initiating Event | |
| Preventative Actions and Safeguards | |
| Contingency Plan and Mitigating Actions | |
| Lessons Learned | |

## Calculations

The following calculations connect safety considerations within this module to knowledge learned in this course to help understand how your knowledge can minimize safety issues.

1. Plant operators saw a vapor cloud (saturated with gasoline vapor) that was 3 ft. high and covered 107 acres. Determine the concentration in mole % of gasoline in the cloud surrounding the tank. How many moles of gasoline were in the vapor cloud? Note: assume the gasoline is pure octane.

2. The upper and lower flammability limits for gasoline are LFL = 1.4 mole % and UFL = 7.6 mole %. Was the cloud gasoline concentration within these limits? Explain.

3. How much energy was released by the CAPECO explosion? How much TNT (in US Tons) would be needed to release an equivalent amount of energy given that the molar enthalpy of TNT is -3407 kJ/mol (NIST Chemistry WebBook, Trinitrotoluene)?

# Advanced Process Safety Modules

The next parts are based on industry practices used to assess process safety and are designed to be used *in upper-level courses*. For professors interested in assigning these parts now, tutorials for both can be found on the University of Michigan SAFEChE website.

## Hazard and Operability (HAZOP) Study

A HAZOP study is a structured analysis of process design to identify potential vulnerabilities in a facility. Review the HAZOP tutorial before completing one for the following system. It is important to note that not all guidewords and parameters will be relevant for different systems. Some information is given here for guidance:

*System to consider*: Secondary Gasoline Storage Tank 409

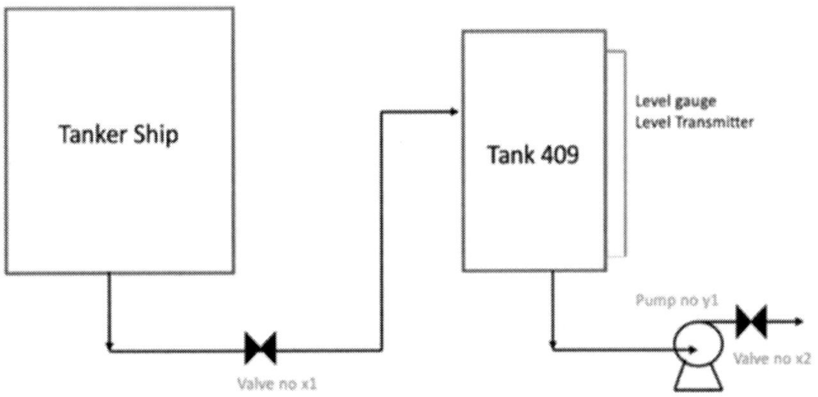


*Process parameters to consider*: Flow, Level

1. Fill out the HAZOP chart below. The first row has been filled out here for you as an example.

| Guideword + Parameter = Deviation | Causes | Consequences | Safeguards | Recommendations |
|---|---|---|---|---|
| **No** Inlet Flow to Tank 409 | Example: Inlet line valve (x1) full closure due to human error | Example: No flow to Tank 409<br>• Low level in tank<br>• pump (y1) damage<br>• release of flammable material | Example: Standard Operating Procedure | Example:<br>1. Low level alarm<br>2. Pump trip on low |
| **More** Inlet Flow to Tank 409 | | | | |
| **Less** Inlet Flow to Tank 409 | | | | |
| **More** Level in Tank 409 | | | | |

## Layers of Protection Analysis (LOPA)

A Layers of Protection Analysis (LOPA) is a semi-quantitative study to identify available safeguards and determine if the safeguards sufficiently protect against a given risk. Review the LOPA tutorial before filling the table out for the system described in this module. Some information is given for guidance:

- Assume that the plant can only accept a moderate risk
- Business losses were estimated to be more than $500 Million

| LOPA Study for CAPECO Explosion | | |
|---|---|---|
| Initiating Event | Cause: | Operator error leading to miscalculated fill time |
| | Consequence: | Gasoline tank overfill leading to vapor cloud explosion |
| | Frequency of Initiating Event (FOIE): | |
| Independent Protection Layers (IPLs) | Description of $IPL_1$, $IPL_2$ ... | $IPL_1$- Physical Containment (Dike) |
| | Probability of Failure of IPLs: PFD = PFD1 x PFD2 x ... | |
| Mitigated consequence frequency (MCF) | MCF = FOIE x PFD | |
| | Category of MCF: | |
| Severity | Impact: | Business losses of more than $500 million |
| | Category: | |
| Risk | Type of risk: | |
| | Acceptable / Unacceptable? | |

### History of the Elective Process Safety Course

For students interested in learning more about Process Safety, the elective course provides additional information and demonstrates industrial application of the concepts. Development of an elective undergraduate Process Safety course was begun in 2015 by Dr. Henry Wang and consisted of a series of lectures by industry experts. This developmental course started as one credit-hour consisting of one two-hour lecture per week following by class discussion. The target audience was undergraduate Chemical Engineering seniors.

In 2016, the course was expanded and offered to graduate students as a two credit-hour course for students completing a term project related to a Process Safety study. In 2017, Kenneth First (recently retired Process Safety Leader at the Dow Chemical Company) became part-time University of Michigan staff and asked to develop a permanent 2 credit-hour Process Safety course. The general direction was development of a course that:

- ✓ Aligned closely with industrial Process Safety training and emphasized "process" relative to "personal" or "occupational" safety.
- ✓ Provides a balance between qualitative and simplified quantitative methods in Chemical Process Risk Analysis.
- ✓ Recognizes that students would have little industrial experience.
- ✓ Builds upon undergraduate Chemical Engineering courses such as Material and Energy Balance, Mass and Heat Transfer, Reaction Kinetics and Thermodynamics.
- ✓ Focuses on practical application of Process Safety concepts such as performing a Hazard Identification and Risk Analysis study.

Student feedback consistently asked that course content be directly related to application in the "real world". It was also recognized that students with little industry experience would not be able to use "experience and engineering judgment" to make Process Safety decisions such that simple quantitative methods (although not highly accurate) might help.

## Alignment of the Elective Course with CCPS Content

It was felt that starting with an "Industrial" Process Safety training program would ensure a focus on practical application. Process Safety training had been developed at the Dow Chemical Company and donated to AIChE's Center for Chemical Process Safety (CCPS). Nearly 1000 industry Chemical Engineers have attended an industry Chemical Hazard Engineering Fundamentals (CHEF) workshop over the past 10 years either in-person or via a webinar format and feedback has been very positive.

- CHEF is organized by "workflow" in performing a Hazard Identification and Risk Analysis study. It had been utilized successfully within Dow for several years and was being updated by CCPS for use across the chemical industry.
- CHEF initially focused on Consequence Analysis, so content has been added to the CCPS version to better accommodate undergraduates including background information, qualitative Hazard Evaluation techniques, Layers of Protection Analysis, and an overview of the 20 elements of CCPS Risk-Based Process Safety.
- AIChE Safety and Chemical Engineering Education (SAChE) eLearning modules were added as homework to supplement CHEF content.

## Elective Process Safety Course Details

The course follows the steps in an overall workflow for performing a Hazard Identification and Risk Analysis (HIRA) study These steps are described in the CCPS CHEF Guide and used in industry to address the following questions:

What is Process Safety?
What are Process Hazards?
What Could Go Wrong?
How Bad Could It Be?
How Likely is It to Happen?
Is the Risk Tolerable?
How can we Sustain and Improve Performance?

*Figure 3 – Chemical Hazard Engineering Fundamentals (CHEF) Guide*

The course incorporates 22 one-hour lectures based on the workflow described in the CHEF Guidebook. There are also small group exercises, various CSB video clips, and larger classroom exercises (group HAZOP and identification of failed Management Systems) in addition to Safety and Chemical Engineering Education (SAChE) eLearning courses (assigned as homework). Finally, a term project including a written report and oral presentation is incorporated (rather than a final exam) to emphasize implementation of concepts.

The course is currently organized as one 2-hour lecture class weekly that covers two topics. Each topic is builds upon concepts from previous lectures, such that the order in which the lectures are presented become important. Course details include:

What is Process Safety?

When we think of safety, we are typically thinking of personal safety – preventing slips, falls, cuts, and burns. We might also think about Industrial Hygiene – preventing personal injuries related to the workplace environment – ergonomics, noise abatement, and chemical exposure. Process Safety, however, is focused on preventing infrequent episodic events such as fines, explosions, and major chemical releases.

Week 1

***Lecture 1: Introduction and Historical Incidents***. In addition to personal injury, Process Safety incidents have the potential to cause significant harm to the public and damage to the environment, occupied buildings, or process equipment. The concepts employed to manage process risk are different from those to manage personal safety.

***Lecture 2: Compliance, Guidance and Metrics.*** Compliance includes government regulations, industry codes and practices, as well as voluntary programs. Performance metrics include Leading and Lagging Indicators.

***Homework**: OSHA Academy 736 Introduction to Process Safety Management (9 modules, 2.5 hours), read Chapter 1-2 of the CHEF Guide, and complete Problem Set.*

What are Process Hazards?

Hazard refers to an inherent property of a substance that can cause an adverse effect. Hazards include chemical hazards such as flammability, toxicity, and reactivity in addition to hazardous process conditions and equipment related hazards. Compiling hazard information is the first step to understanding hazards. There are several excellent resources to find chemical hazard information in preparation for a Hazard Identification and Risk Analysis study. Safety Data Sheets (SDS) are a typical source of information for hazards associated with the chemicals. Information for flammability, toxicity, and environmental issues are commonly included.

Week 2

***Lecture 3: Flammability Hazards.*** Losses attributed to fires at US industrial and manufacturing sites averaged $1.2 billion per year in direct property damage alone from 2011 to 2015 (excluding business loss and costs associated with injuries or deaths). Potential consequences from fires within chemical and hydrocarbon facilities can be significant. Understanding key parameters associated with flammability such as flash point, flammable limits, limiting oxidant concentration, and minimum ignition energy is important to determine the conditions under which fire (and explosion) may occur.

***Lecture 4: Toxicity Hazards***. Toxicity is the quality, state, or degree to which a substance is poisonous and/or may chemically produce an injurious or deadly effect upon introduction into a living organism. Process Safety focuses on exposure associated with a single episodic event by addressing acute toxic impacts for both employees and the public. Inhalation and dermal adsorption are the most significant routes in Process Safety studies. However, Ingestion Toxicity may be important to understand potential environmental impacts of releases or contamination of the food chain.

***Homework:** SAChE ELA 951 Hazard Recognition (3 Units, 2 hours), read Chapter 3-4 of the CHEF Guide, ,and complete Problem Set.*

Week 3

**Lecture 5: Reactivity Hazards**. Reactive hazards are the dangers associated with uncontrolled chemical reactions in industrial processes. These uncontrolled reactions - such as thermal runaways and chemical decompositions - have been responsible for numerous fires, explosions, and toxic gas releases. Evaluation of reactivity hazards often involve interpretation of reactive chemicals test data. Endothermic reactions represent a potential hazard if gaseous or highly volatile products are generated. Exothermic reactions have the potential for a runaway reaction leading to a dramatic increase in temperature, pressure (if the reaction is contained) and reaction rate. Raw materials, process streams, products and waste of any process should be reviewed and evaluated to determine if any potential reactive chemical hazards are involved.

**Lecture 6: Other Hazards.** The concept of various types of energy and a relative state for comparing severity may be helpful to identify other process situations where Hazards may exist.

***Homework:** SAChE ELA 962 Reactivity Hazards (3 Units, 2 hours), read Chapter 5-6 of the CHEF Guide and complete Problem Set.*

What Can Go Wrong?

In managing process hazards and risks, an engineer needs to *anticipate* process upsets or equipment failures (including control equipment) that could result in a Loss Event. Loss Events include loss of quality, loss of productivity, and loss of containment. Process Safety focuses on loss of containment events although minimizing other types of loss events is equally important in Chemical Engineering.

Week 4

***Lecture 7: Learning From Experience.*** Key characteristics for processing equipment, combined with historical incident data, provides helpful information related to "what can go wrong?" Experience and Historical Events may be related to specific types of equipment including Fluid Transfer Equipment, Heat Transfer Equipment, Rotating Equipment, Reactors, Fired Equipment, Chemical Storage, and Solids Handling Equipment.

***Lecture 8: Inherently Safer Design***. The concept is changes to the process that either eliminates the hazard completely or sufficiently reduces its magnitude or likelihood of occurrence to minimize the need for elaborate safety systems and procedures. Furthermore, this hazard elimination or reduction would be accomplished by means that were inherent in the process, and, thus, permanent and inseparable from it. An option may be inherently safer regarding some hazards but may be inherently less safe regarding others, decisions about the optimum strategy for managing risks from all hazards are required. The decision process must consider the entire life cycle, the full spectrum of hazards and risks, and the potential for transfer of risk from one impacted population to another.

***Homework:*** *SAChE ELA 984 Inherently Safer Design (3 Units, 2 hours), read Chapter 7-8 of the CHEF Guide, and complete Problem Set.*

Week 5

**Lecture 9: Hazard Evaluation Techniques.** Hazard Evaluation is an organized effort to identify and analyze the significance of hazardous situations associated with a process or activity. To meet evaluation requirements, different techniques (such as Checklists, What If Analysis, Failure Modes and Effects Analysis, or Hazard and Operability Study) might be used for portions of the process with the greatest hazards. Generally, hazard severity and consequence severity are related with high hazard typically leading to high severity consequences.

**Lecture 10: Hazard and Operability Studies.** HAZOP is a commonly used systematic qualitative technique to identify process hazards and potential operating problems using a series of guide words to study process deviations. It is used to question every part of a process to discover what deviations from the intention of the design can occur and what their causes and consequences may be. This is done systematically by applying suitable guide words. This is a systematic detailed review technique, for both batch and continuous plants, which can be applied to new or existing processes to identify hazards.

***Homework:*** *SAChE ELA 970 What Can Go Wrong? (4 Units, 3 hours), read Chapter 9-10 of the CHEF Guide and complete Problem Set*

Week 6

***Lecture 11: Development of Hazard Scenarios***. Development of a good scenario statement will help tremendously understanding of the sequence of events and in performing the risk analysis. A Scenario represents a detailed description of an unplanned event or incident sequence that results in a loss event and its associated impacts, including the success or failure of safeguards involved in the incident sequence. A good scenario description is needed to ensure all team members fully understand the event sequence with the associated Initiating Event, Loss Event, Outcome and

Consequences. This understanding is key to identifying cost-effective Safeguards and Protective Layers.

***Hazard Evaluation and Scenario Development Class Exercise***

***Homework:*** *SAChE ELA 974 Hazard Identification and Risk Analysis (2 Units, 2.5 hours), read Chapter 11 of the CHEF Guide and complete Problem Set*

How Bad Could it Be?

Consequence Analysis if often done qualitatively as part of Hazard Identification. When additional information is desired to make risk Management decisions, a more quantitative approach may be performed that involves source models, dispersions models, explosion models and impact or effect models. Models used for Consequence Analysis have very high uncertainty such that results may be significantly worse or better when estimating the consequence of actual historical incidents. Nearly all models are based on some integral of a mass, volume, momentum, or energy balance.

*Week 7*

***Lecture 12: Source Models.*** Generally, a vessel or other process equipment within a chemical facility has feeds such as a liquid or vapor and outlets. In addition, during a process upset, material may discharge from an emergency pressure relief device or leak from failed nozzles or piping. Source models are used to estimate discharge rate, total quantity released, and extent of flash and/or evaporation for liquids or aerosol droplets.

***Lecture 13: Vapor Dispersions.*** Vapor dispersion modeling is estimation of the dissipation of a toxic or flammable vapor cloud in air due to wind, thermal action, gravity spreading, and atmospheric turbulence. It is used to estimate the concentration as a function of distance from the leak source and the area affected in a chemical release.

***Homework:*** *SAChE ELA 965 Source Models (3 Units, 2 hours), read Chapter 12-13 of the CHEF Guide and complete Problem Set*

Week 8

***Lecture 14: Explosions.*** An explosion is a rapid expansion of gases resulting in a rapidly moving pressure or shock wave. The expansion can be mechanical (by means of a sudden rupture of a pressurized vessel) or it can be the result of a rapid chemical reaction. The various types of explosion are related. For Vessel or Equipment Rupture, Physical Explosion is typically related to release of a pressurized gas. A BLEVE occurs when significant vaporization occurs such that the gas created participates in the explosion energy. If a Runaway Reaction occurs very rapidly, a Chemical Explosion may occur. For slower or less energetic reactions, vessel or equipment rupture may result. In combustion related (flammable vapor or combustible dust) explosions may occur within confined spaces or as an outdoor vapor cloud explosion.

***Lecture 15: Impact Assessment.*** Impact Estimation or Effect Modeling is assessment of the consequences for specific incident outcome. For assessing effects on personnel, consequences may be expressed as fatalities or injuries. If physical property or business interruption is the objective, the consequence may be monetary loss. Environment impacts may be much more complex and could include impact on plant or animal life, soil contamination, damage to natural resources, or other impacts. Consequences are often difficult to assess precisely, such that order-of-magnitude categorization is often used.

***Homework:*** *SAChE ELA 967 Atmospheric Dispersion (3 Units, 3.25 hours), read Chapter 14-15 of the CHEF Guide, submit Term Project Progress Report, and complete Problem Set*

How Likely is It to Happen?

Once the consequence has been determined, a measure of likelihood or frequency is needed for an estimate of risk. Risk is a measure of economic loss, human injury, or environmental damage in terms of both the likelihood and the magnitude. In addition to historical data, we often use Fault Tree or Event Tree techniques. These techniques utilize probability theory and Boolean (AND or OR logic) to estimate probability. The estimates are highly uncertain and often only order-of-magnitude values are used.

Week 9

***Lecture 16: Likelihood Evaluation.*** Probability theory is the branch of mathematics concerned with likelihood. Although it is not possible to perfectly predict random events, much can be said about their behavior. Incidents in chemical facilities are often the result of complex interactions of process components. In some cases, this interaction requires the simultaneous failure of several components. This parallel structure is often depicted by a logical "AND" with the failure probability of the individual components multiplied to obtain an overall failure probability. Process components may also interact in series which is often depicted by a logical "OR" with failure probability of the individual components added to obtain an overall failure probability.

***Lecture 17: Risk Analysis and Risk Assessment.*** Risk Analysis involves qualitative and quantitative Hazard Evaluation methods. Qualitative Hazard Evaluation techniques are commonly used to pinpoint weaknesses in process design and facility operation that could lead to incidents with potential safety consequences or impacts. Quantitative methods are typically used to determine to estimate the likelihood of occurrence and the magnitude of an incident for a specific sequence of events leading to an Incident. Risk Assessment in one form or another is inherent in all decision making. If the risk exceeds predefined and agreed upon criteria:

- The activity creating the risk must be reviewed with corporate management to agree upon appropriate risk reduction actions, or
- Methods of reducing the risk must be determined and implemented, or
- The activity creating the risk must be discontinued.

***Homework:*** *SAChE ELA 988 Damage Mechanisms: Asset Integrity and Reliability (4 Units, 3 hours), read Chapter 16-17 of the CHEF Guide, and complete Problem Set*

Is the Risk Tolerable?

In addition to determining risk as a function of consequence severity and frequency, risk is managed by identifying and implementing cost-effective safeguards or protections. Key questions addressed during a risk analysis and assessment include:

- How safe is safe enough?
- How many protection layers are needed?
- How much risk reduction is provided by each protection layer?

Safeguards or protective layers are identified and evaluated for effectiveness and reliability. Safeguards are either *preventive* (reduce the likelihood of the incident occurring) or *mitigating* (reduce the severity of the outcome once a loss event has occurred). For instrumented safeguards or interlocks, students determine the necessary reliability (or probability of failure on demand) necessary to meet the tolerable risk criteria.

Week 10

***Lecture 18: Layers of Protection Analysis.*** Layers of Protection Analysis is a Risk Assessment approach used to compare a scenario risk estimate to risk criteria for determining where additional risk reduction or more detailed analysis is needed. It is a simplified method of risk assessment that provides the middle ground between a qualitative process hazard analysis and a traditional, expensive quantitative risk analysis. The approach analyzes one incident scenario (cause-consequence pair) at a time, using simplifying rules to evaluate initiating event frequency, independent layers of protection failure probabilities, and consequence severity to provide an order-of-magnitude estimate of risk that may be compared to a company's tolerable risk criteria.
The primary purpose of LOPA is to determine if there are sufficient layers of protection against an incident scenario.

***Lecture 19: Procedures and Human Performance.*** Operating and maintenance procedures are essential to achieving proper training and safe, efficient performance of operating and maintenance tasks. Intended or unintended human action or inaction that produces an inappropriate result is a common Initiating Event for many catastrophic incidents. On the other hand, proper execution of an emergency procedure is a common Safeguard or Protective Layer. Human Performance by designers, operators, mechanics, electricians, engineers, or managers may contribute to or result in process safety incidents.

***Homework:*** *SAChE ELA 980 Risk Review Using LOPA (3 Units, 2.5 hours). read Chapter 18-19 of the CHEF Guide and complete Problem Set*

Week 11

***Lecture 20: Preventive Safeguards.*** A preventive barrier acts between the initiating event (the cause) and a loss event, helping reduce the frequency of the incident scenario, and thus, reducing the scenario's risk.

***Lecture 21: Mitigating Safeguards.*** A mitigating safeguard acts once the loss event has occurred and is intended to eliminate the consequence or significantly reduce the consequence severity, thereby reducing the scenario's risk. As with preventive Safeguards, there are Passive, Active and Procedural mitigating safeguards. Often the passive safeguards are more reliable that active or procedural. Procedural safeguards, such as Emergency Response (although very important) are generally not sufficiently reliable to be considered Protection Layers.

***Homework:*** *SAChE ELA 997 Risk Based Process Safety - Manage Risk (4 Units, 3 hours), read Chapter 20-22 of the CHEF Guide and complete Problem Set.*

<u>How Can We Sustain and Improve Performance?</u>

In Process Safety, the Management Systems help to ensure the gains from Hazard Identification and Risk Analysis and Implementation of Safeguards or Protective Layers are sustained over time. A Management System is a formally established and documented set of activities designed to produce specific results in a consistent manner on a sustainable basis. History has shown us that most major incidents could have been prevented if all management systems were effective and functioning.

Week 12

***Lecture 22: Management Systems and Risk Based Process Safety.*** The purpose of these Risk Based Process Safety Guidelines is to help organizations design and implement more effective process safety management systems. The Risk Based Process Safety approach recognizes that all hazards and risks are not equal; consequently, it advocates that more resources should be focused on more significant hazards and higher risks. The approach is built on four foundational blocks:

- Commit to Process Safety
- Understand Hazards and Risk
- Manage Risk
- Learn from Experience

***Management Systems Class Exercise***

***Homework:*** *Term Project*

Week 13

**Submit Term Project Reports** - *All*

**Term Project Oral Presentations** – *All*

**Class Feedback**– All

Various software is utilized in the course including the Chemical Reactivity Worksheet (CRW 4 available from AIChE), CAMEO Chemicals, ALOHA Vapor Dispersion Modeling, and the CHEF Calculation Aid (available from AIChE/CCPS). Each of these software tools is available as a free download.

*Figure 4 – CRW Software*

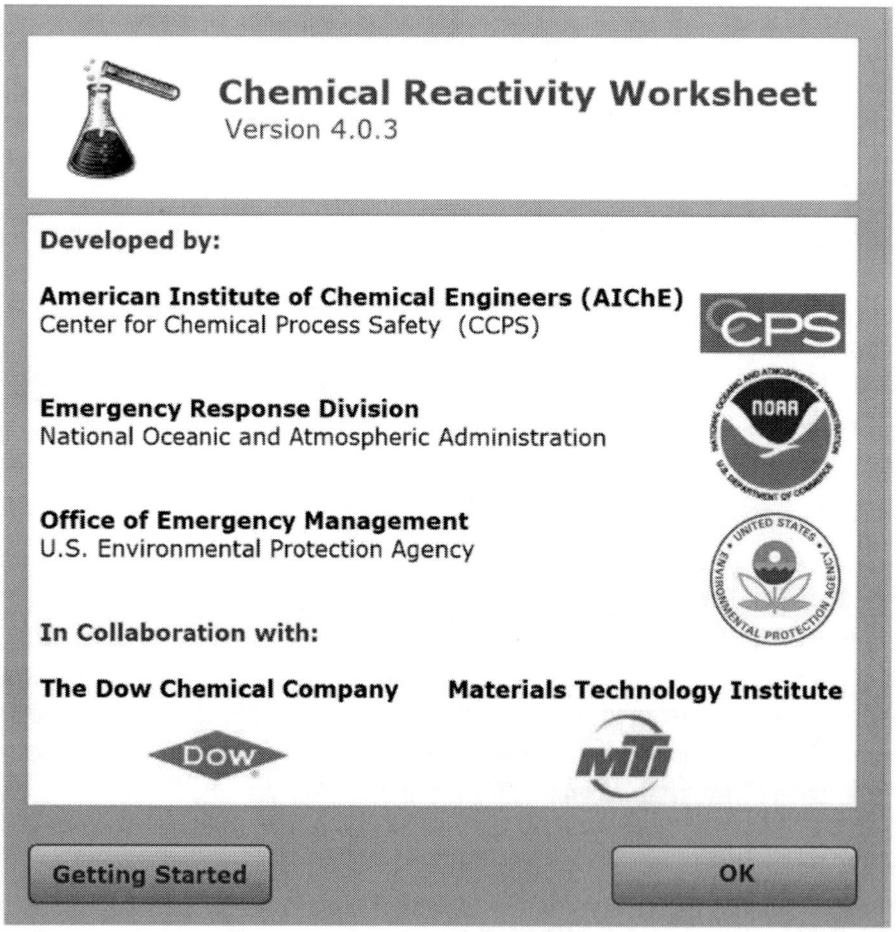

*Figure 5 – CAMEO Software*

# CAMEO *Chemicals*

## Database of Hazardous Materials

Search

Find response information for thousands of hazardous materials, including fire and explosion hazards, health hazards, firefighting techniques, cleanup procedures, protective clothing, and chemical properties.

MyChemicals

Build a list of chemicals. For example, substances involved in an incident response (such as a train derailment) or chemicals stored in your community.

Reactivity

See what hazards might occur if chemicals in your MyChemicals collection are mixed together.

*Figure 6 –ALOHA Software*

# ALOHA Software

ALOHA® (Areal Locations of Hazardous Atmospheres) is the hazard modeling program for the CAMEO® software suite (Computer-Aided Management of Emergency Operations), which is used widely to plan for and respond to chemical emergencies.

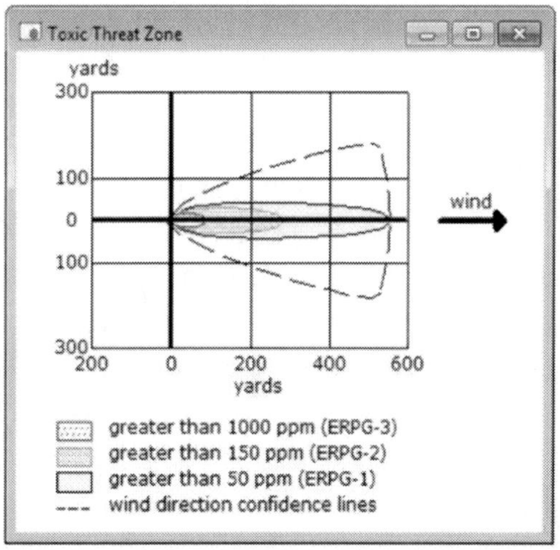


ALOHA allows you to enter details about a real or potential chemical release, and then it will generate threat zone estimates for various types of hazards. ALOHA can model toxic gas clouds, flammable gas clouds, BLEVEs (Boiling Liquid Expanding Vapor Explosions), jet fires, pool fires, and vapor cloud explosions. The threat zone estimates are shown on a grid in ALOHA, and they can also be plotted on maps in MARPLOT® (Mapping Application for Response, Planning, and Local Operational Tasks), Esri's ArcMap, Google Earth, and Google Maps. The red threat zone represents the worst hazard level, and the orange and yellow threat zones represent areas of decreasing hazard.

*Figure 7 – CHEF Calculation Aid*

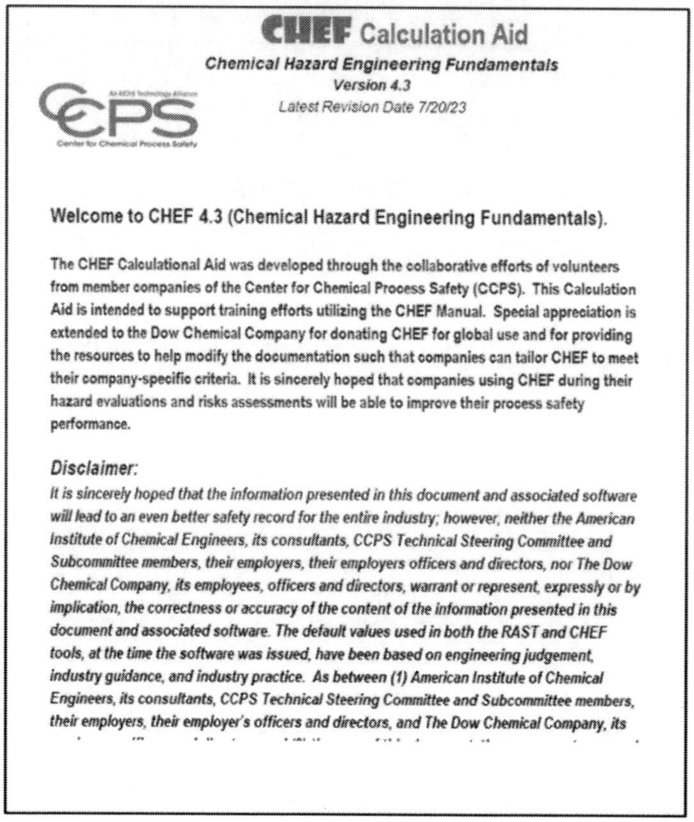

**CHEF Calculation Aid**

***Chemical Hazard Engineering Fundamentals***

***Version 4.3***

*Latest Revision Date 7/20/23*

CCPS

Center for Chemical Process Safety

**Welcome to CHEF 4.3 (Chemical Hazard Engineering Fundamentals).**

The CHEF Calculational Aid was developed through the collaborative efforts of volunteers from member companies of the Center for Chemical Process Safety (CCPS). This Calculation Aid is intended to support training efforts utilizing the CHEF Manual. Special appreciation is extended to the Dow Chemical Company for donating CHEF for global use and for providing the resources to help modify the documentation such that companies can tailor CHEF to meet their company-specific criteria. It is sincerely hoped that companies using CHEF during their hazard evaluations and risks assessments will be able to improve their process safety performance.

***Disclaimer:***

*It is sincerely hoped that the information presented in this document and associated software will lead to an even better safety record for the entire industry; however, neither the American Institute of Chemical Engineers, its consultants, CCPS Technical Steering Committee and Subcommittee members, their employers, their employers officers and directors, nor The Dow Chemical Company, its employees, officers and directors, warrant or represent, expressly or by implication, the correctness or accuracy of the content of the information presented in this document and associated software. The default values used in both the RAST and CHEF tools, at the time the software was issued, have been based on engineering judgement, industry guidance, and industry practice. As between (1) American Institute of Chemical Engineers, its consultants, CCPS Technical Steering Committee and Subcommittee members, their employers, their employer's officers and directors, and The Dow Chemical Company, its*

The course format, group exercises, and lecture materials have been used for the past 5 years and updated based on student and "industry advisor" feedback. As new SAChE eLearning courses become available, each is assessed to determine if it might better supplement CHEF content and replace a course currently used. For 2023, an OSHA Academy introduction to Process Safety Management course and 10 SAChE courses were utilized including:

Level 1 - ELA 951 Hazard Recognition
Level 2 - ELA 962 Reactivity Hazards
ELA 965 Source Models
ELA 967 Atmospheric Dispersion
ELA 970 What Can Go Wrong?
ELA 974 Hazard Identification and Risk Analysis
Level 3 - ELA 980 Risk Review using Layers Of Protection Analysis
ELA 984 Inherently Safer Design
ELA 988 Damage Mechanisms: Asset Integrity and Reliability
ELA 997 Risk Based Process Safety: Manage Risk

## Future Plans

Looking forward, only 10-15% of the Chemical Engineering students participate in ChE 407. We are currently looking at how we might increase participation. We are also working on a succession plan to ensure that "qualified" instructors and industry experts are available for the course to continue well into the future. Finally, we are looking into how this course might be expanded to three credit-hours by adding next lectures, expanding group exercises, and adding SAChE eLearning courses.

# 54g Instruction of Process Safety Engineering through Project Based Learning (PBL)

*Pedro J. Arias-Monje, Sepehr Arbabi*

To "hold paramount the safety, health and welfare of the public and protect the environment in performance of their professional duties" is the first mandate in the Code of Ethics of chemical engineers. We have found the Process Safety Engineering course to be a perfect setting to instruct our students on this tenet, particularly when implementing Project Based Learning (PBL) and using multimedia generated by the U.S. Chemical Safety and Hazard Investigation Board, in addition to traditional textbook material (Crowl and Louvar, 4$^{th}$ Edition). This methodology has been used with small classes (up to 8 students), and sometimes with students from other engineering fields (e.g.; mechanical and electrical engineering) and developed over two years. Examples of semester-long academic projects that fully engage the students and incorporate principles of hazard identification, industrial hygiene, risk assessment and safety design will be provided. We have found that introduction of complex design problems outside senior design or capstone courses can be initially overwhelming to students, but after providing project management tools (like task breakdown and strategic planning), the students readily engage with the topic, learn while working on the project and perform well on other course assignments and exams. Finally, we discuss how the tragic nature of obliquity of industrial and chemical accidents, presents an opportunity for the continuous improvement of a Process Safety Engineering class.

## 54h On the Use of CSB Videos for Reflective Learning in Teaching Process Safety

*Paul Amyotte*

The objective of this presentation is to demonstrate the pedagogical value of videos produced by the United States Chemical Safety and Hazard Investigation Board (US Chemical Safety Board or CSB) in teaching process safety to chemical engineering undergraduate students. Drawing on the author's experience over the past 20 years teaching process safety at Dalhousie University (Amyotte, 2013), the usefulness of these videos in a reflective learning mode is described.

As indicated on its web site (*www.csb.gov*), the CSB is an independent, nonregulatory federal agency that investigates the root causes of major chemical incidents. The CSB is a valuable source of process safety information and case studies in both written and video formats (Amyotte and Khan, 2021). CSB videos have been recognized globally for their educational value. They are also free to access and use.

Noted chemical engineering educator Richard Felder has commented on the importance of *teaching to all types* (Felder, 1996). It is important to provide opportunities for both reflection (thinking) about course material and active participation in discussing course concepts (Felder, 1996). While there has been a welcome shift in recent years to increased emphasis on active learning, the coronavirus pandemic has necessitated a return, in part, to reflective learning via online teaching. A side-benefit for the current author has been the recognition that while active learning is indeed an engaging mode of course delivery, reflection can also be highly beneficial in helping students learn.

For the past three years, the author has allocated 20% of the overall process safety course grade to a series of ten video reflections (200 words each). Students briefly describe the video itself and then reflect on the key takeaway lessons for them personally. Videos are, for the most part, drawn from the CSB video room (*www.csb.gov*) and are viewed together in a classroom setting. Selection of the video for reflection is left to the student's choice from among the videos watched in a given course module:

- 1. Introduction to Process Safety
    - *Back to School Safety Message* (CSB)
    - *After the Rainbow* (CSB)
    - *Preliminary Animation of Philadelphia Energy Solutions Refinery Fire* (CSB)
    - *Wake Up Call: Refinery Disaster in Philadelphia* (CSB)
- 2. Loss Causation
    - *Chevron Richmond Refinery Fire Animation* (CSB)
- 3. Process Safety Management
    - *Process Safety: Who's Responsible?* (Yut Art)
    - *Courage for Safety* (Woodside)
- 4. Incident Investigation/Hazard Identification
    - *Mixed Connection, Toxic Result* (CSB)
    - *Hot Work: Hidden Hazards* (CSB)
- 5. Hazard Identification/Practices and Procedures
    - *No Escape: Dangers of Confined Spaces* (CSB)
    - *Spiral to Disaster* (BBC)
- 6. Communications/Training/Emergency Preparedness
    - *Hazards of Nitrogen Asphyxiation* (CSB)

  - *Lockout/Tagout* (Coastal)
  - *Emergency Response Safety Message* (CSB)
- 7. Inherently Safer Design
  - *Inherently Safer: The Future of Risk Reduction* (CSB)
  - *Inherently Safer Process Design* (IChemE)
- 8. Fires and Explosions
  - *Filling Blind* (CSB)
  - *Deadly Contract* (CSB)
  - *Inferno: Dust Explosion at Imperial Sugar* (CSB)
  - *Explosion Protection* (ESCIS)
- 9. Risk Assessment and Management
  - *Process Safety: Managing Risk in High-Hazard Sectors* (WorkSafeBC)
  - *Uncovered Hazards: Explosion at the DeRidder Pulp and Paper Mill* (CSB)
- 10. Case Studies
  - *One Night in Bhopal* (BBC)
  - *Anatomy of a Disaster* (CSB)

Student feedback on the video reflections, as well as examples of their work, will be given during the presentation. Conclusions will also be made on the use of visual media in a process safety educational environment. Overall, the experience has been highly positive from both student and instructor perspectives.

**References**

Amyotte, P.R., *Process Safety Educational Determinants*, Process Safety Progress, **32**, 126-130 (2013).

Amyotte, P.R. and Khan, F.I., *The Role of Inherently Safer Design in Process Safety*, Canadian Journal of Chemical Engineering, **99**, 853-871 (2021).

Felder, R.M., *Matters of Style*, ASEE Prism, **6** (4), 18-23 (1996).

## 42a Electrically Heated Equipment Safety in Pilot Plants

*Richard Palluzi*

Pilot Plants often use electrically heated equipment for heat tracing, process heating, and similar applications due to its convenience, flexibility, and relatively low cost. Unfortunately, electrically heated equipment creates significant safety issues including:

- Overheating a component raising the potential for a catastrophic component failure under pressure
- Overheating a seal and creating a significant leak
- Providing an ignition source to ignite a leak or release
- Increasing the potential for and probability of a release in standard piping and tubing fittings

This presentation will discuss these potential hazards , the conditions and situations where they are more likely to occur and suggest guidelines to avoid or reduce their likelihood.

## 42d What to Expect When Expecting in Lab: Safety Considerations and Green Chemistry Resources for Pregnant Researchers in the Chemical Laboratory

*Mary Kate Lane, Paul T. Anastas, Julie B. Zimmerman*

The importance of laboratory safety and the awareness of working with laboratory hazards becomes heightened when chemical researchers become pregnant and can lead to uneasiness about lab work. In an effort to retain and support women and parents in chemical careers, we present our recent review of unique risks and resources for pregnant researchers in the chemical laboratory, where a chemical laboratory is defined as an industry or academic laboratory engaging in primary research in fields like chemistry, engineering, or medicine, but excluding biological work with substances such as pathogens or live organisms. We examine lab relevant doses for commonly encountered hazards for pregnant lab workers, including chemical hazards (organic solvents, heavy metals, engineered nanomaterials, and endocrine disruptors), radiation hazards (ionizing radiation producing equipment and materials and non-ionizing radiation producing equipment), and other hazards related to the lab environment (excessive noise, excessive heat, psychosocial stress, strenuous physical work, and/or abnormal working hours). Inspired by the principles of Green Chemistry and Green Engineering, reducing hazard is the most promising route to eliminating or at least minimizing risk—as it is an intrinsic and constant property of chemicals and processes—whereas exposure controls can and often fail. Therefore, implementing green chemistry could directly benefit pregnant researchers in a laboratory setting. This oral presentation will share the main findings of this review, highlight many green chemistry tools, and indicate where green chemistry can aid in furthering creating safe laboratory spaces for pregnant researchers.

## 42b Evaluation of Filler Containing Si-H (Silicon hydride) Formulations for $H_2$ off-Gassing Risks during Production and Packaging

*Hulusi Turgut, Jeff Kelly, Nate Stelzer, Craig Gorin, Chi-Hao Chang, Rocky Zhu, Travis Scholtz, Ryan Gilliland, Michael Whitbrodt, Jyo Lyn Hor, Zach Rolfs*

Silicon compounds containing silicon-hydrogen bonds including chlorosilanes, Si-H silicone fluids and Si-H emulsions play a key role in a wide range of applications from aerospace to textiles. Numerous types of fillers are added into the Si-H containing products to gain added functionality. When fillers are added into the Si-H containing formulation, it can cause Si-H compounds to liberate $H_2$ due to basicity of the filler. $H_2$ liberation could present a major risk from drum bulging to explosion or fire. Understanding how to evaluate filler Si-H interaction and paying detailed attention to equipment design, operating procedures, storage and shipping conditions are critical to mitigate the $H_2$ off-gassing associated risks.

In this study, commonly used fillers were evaluated in Si-H containing formulations using thermal screening unit (TSU) and gas chromatography (GC). The filler pH level was measured as received and compared with the manufacturer technical data sheet. The study investigated various scenarios including temperature ramp up during shipping, temperature increase during production and packaging. The results showed that 1 kg of finished Si-H containing product could liberate up to ~0.02 mole of $H_2$ during the first 10 hours after filler is added into the Si-H system. $H_2$ off-gassing could also cause thicker products to swell by as much as 30% of its original volume which could potentially cause production or packaging vessel overflow.

The key finding is that fillers with alkaline impurities could trigger unintended $H_2$ evolution from the Si-H containing compounds. Controls such as passivating production vessels, installing pressure relief valve device on the packaging line, using vented packaging containers and leaving sufficient headspace in the packaging containers could be used to mitigate the risks.

## 42c Overlooked Considerations during Safety Reviews of Lab Scale Systems and Pilot Plants

*Pratik Bhishikar*

Safety reviews are necessary to ensure the design, fabrication, testing, and installation of process plants complies with safety, operability, health, and environmental requirements and standards. Safety reviews are conducted for plants of different scales (lab, pilot, demonstration, and production scales) throughout the life of the project. Compared to demonstration and production plants, lab scale systems and pilot plants are relatively small in size, with much lower volumetric throughput, making it difficult to consistently apply commercial health and safety standards. Furthermore, a few areas are commonly overlooked during safety reviews of lab scale systems and pilot plants, which can potentially lead to increased risk, or a process hazard. This article highlights these areas, where additional attention applied during the health and safety review process will pay dividends for these small scale plants.

## 42e Utilizing Simple Heat Transfer Models for Safe Scale-Ups of Exothermic Chemistries

*Seshasayee Mahadevan, Jessica Nichols, Jason S. Fisk, Mike DePierro, Jeff Kelly*

Sophisticated modeling of energy release in reactors can require extensive information about the chemistry and physical properties of the system. By contrast, a typical hazard evaluation may generate very good information about theoretical worst-case energy releases but may provide limited information regarding chemical kinetics-based energetics. The generation and validation of formal kinetic models for new chemistries may be a time-intensive process as well. Thus, development of some simple heat transfer models to predict temperature distribution in lab-scale reactors of varying reactor sizes, fill levels, and chemistries are illustrated.

Hazard evaluation testing that quantifies energy release, such as dynamic or adiabatic calorimetry, provides critical process safety information for desired chemistries but may not capture desired kinetics. Using simple heat transfer models, the energy release values from hazard evaluation testing are integrated with heat transfer models in COMSOL Multiphysics to predict temperature distribution within reactor set ups. These are further integrated with experimental observations for well-known chemistries to ascertain model validity.

This talk will highlight how this approach can generate insights into the safe scale-up and operation of exothermic reactions in lab-scale reactors. The talk also illustrates the application of this method to a sample scale-up problem and also compares outcomes to results from more sophisticated thermokinetic models and simulations. While the methodology employed herein is not a substitute for a thorough hazard evaluation, it can provide a scientific method to aid in reactor selection and experimental design.

## 42f Utilizing a Digital Process Safety Game to Improve Training Strategies in Pilot Plants

*Brittany Butler, Cayla Ritz, Jeffrey Stransky, Cheryl A. Bodnar, Emily Dringenberg, Elif Miskioglu*

Aside from the development of new process technologies, laboratories and pilot plants cultivate an environment where improved training practices can be implemented and refined, due to the high level of safety standards required in pilot scale plants. Current training methods used to manage process safety risks include process hazard analyses (such as operability studies and what-if checklists), and risk assessments (such as hazard and operability (HAZOP) and layers of protection analysis (LOPA)). These analysis methods rely on the use of engineering judgement and expert opinion, however, these analyses are not free from the influence of heuristics and cognitive biases that come along with human judgements (Baybutt, 2018). Biases can lead to blind spots when making judgements, where an individual's beliefs about how they will react does not reflect the behaviors that actually take place (Bazerman & Tenbrunsel, 2011). When our beliefs and behaviors are not in alignment, our resulting behavior may be uninformed to potential consequences. Blind spots can also be caused by the lack of recognition of external pressures, and how they impact judgements, sometimes contributing to the occurrence of process safety incidents.

External pressures, such as productivity, money, and reputation, can impact judgements that result in process safety incidents. As described by the Chemical Safety Board (2023), there have been several instances that reflect these types of occurrences. For instance, in 2014, Freedom Industries chemical storage and distribution leaked crude MCHM and polyglycol ethers into a river that impacted drinking water for the surrounding nine counties. The CSB investigation found no documentation of inspections or maintenance on the tank that leaked, and that the secondary containment surrounding the tank had been deteriorated. While the company knew the costs to repair these items, they had not moved forward on a project to make the changes. Another example is from the Chevron refinery fire in Richmond, California that resulted in a chemical leak which eventually ignited to a fireball, injuring six employees and causing 15,000 people in the surrounding area to seek medical treatment. During the investigation, the CSB found that the leak originated from a section of an insulated pipe that had been flagged twice for maintenance. The plant was aware of this, but continued to run the process until they could pinpoint the leak, which eventually resulted in the chemical release, and consequent explosion (CSB, 2023). Incidents like these suggest that other criteria such as money, down-time, and production are influencing process safety judgements, and are taking precedence over safety.

Previous work suggests there are six competing criteria that may impact process safety judgements. These six criteria are: safety, time, production, spending, leadership, and relationships. For example, in the Freedom Industries leak, safety was sacrificed for the cost to repair the secondary containment, and the time to complete the inspections and maintenance on the tanks. Whereas, the Chevron Refinery fire shows the prioritization of production and time over safety, by not repairing the leak on the pipe which resulted in the fire and explosion. For these reasons, it is important that industry professionals are aware of competing criteria when approaching process safety judgements.

In order to study how these competing criteria compare to one another when making process safety judgements, we use the game-based learning environment Contents Under Pressure (CUP). CUP is a professionally developed game where participants play the role of a chemical plant manager facing an impending hurricane. Within the game, participants are able to make authentic safety judgements, without the associated risks. Over the course of 15 simulated game days, participants make a series of process safety judgements which impact metrics such as time, safety, reputation, and plant output.

Research indicates that a way to mitigate poor judgements could be to engage individuals in self-reflection about any gaps that exist between their beliefs and performed behaviors. A pilot study was conducted in spring 2023 to determine the differences that may exist between the beliefs and behaviors of industry professionals when approaching process safety judgements. Five industry professionals with engineering backgrounds who work in the chemical process industry were recruited for the study. The methodological approach for this study collected practitioner beliefs (pre-CUP interview), simulated behaviors (CUP), and reflection (post-CUP survey) through ranking the competing criteria from the conceptual framework. This allowed researchers to directly compare any gaps that existed between participant beliefs and behaviors.

We anticipate this work will help raise awareness of approaches to process safety judgements and will also generate conversation on how to build on current training approaches to include consideration for the impacts of competing criteria on process safety judgements. Implementing new or improved training strategies in laboratories and pilot plants can assist in refining these strategies, before rolling them out to larger production plants. This increase in awareness can help mitigate the occurrence of process safety incidents related to engineering judgement in the future.

# 85a Internally Heat-Integrated Pressure-Swing Distillation for Non-Ideal Separation Using Computational Fluid Dynamics

*Heecheon Lee, Chaeyeong Seo, Minyong Lee, Jae Lee*

This study presents the computational fluid dynamics (CFD) simulation to design internally heat-integrated pressure-swing distillation (HIPSD) systems with improved energy efficiency for azeotropic distillation. A pinch analysis was applied to determine the optimal operating pressure of the HIPSD in a double annular column configuration to achieve adequate heat transfer and circumvent distillation boundaries. To separate a highly azeotropic ternary mixture of butyl acetate, butanol, and water, a single unit of HIPSD and a decanter are employed. Different design alternatives are explored for various initial feed compositions. In each sequence, the CFD method was used to calculate the heat transfer rate inside the HIPSD, leading to a reduction of total utility consumption by 9.72% and 15.44%. The reboiler and condenser duty in the HIPSD were also reduced by up to 48.65%. Furthermore, the internal heat integration improves the separation efficiency in the condenser of the high-pressure column, allowing for zero reflux.

## 85b An Energy-Efficient Double Annular Column for Azeotropic Separation

*Heecheon Lee, Minyong Lee, Jae Lee*

This study presents a novel approach to design an internally heat-integrated pressure swing distillation (HIPSD) as a double annular column with a pinch-based method. The proposed configuration involves the annular stripping section of a low-pressure (LP) column partially wrapping around the rectifying section of an inner high-pressure (HP) column in order to minimize exergy loss. To achieve adequate heat transfer and azeotropic separation, the pinch pressure is determined by considering the minimum temperature approach and circumvention of the azeotropic boundary. The heat integration between the two columns accounts for the overall differences in heat transfer coefficients for each stage, which are affected by the composition of the fluid and its hydrodynamic characteristics of each stage. Additionally, each column is sized to meet hydraulic conditions for optimal separation efficiency. The HIPSD design for several azeotropic mixtures was designed with maximum energy savings of up to 23.80%.

## 85d Efficient Cleaning Methods for Gas Supply Systems in Semiconductor Manufacture Using CFD

*Ireh Seo, Rakyoung Jeon, Tae-Ung Yoon, Chang-ha Lee*

This study focuses on developing effective cleaning methods for specialty gas supply systems used in the semiconductor manufacturing industry. In this industry, it is crucial to maintain impurity levels within the range of parts per billion (ppb) to parts per million (ppm) for optimal manufacturing efficiency. The gas supply system comprises various components such as filters, diaphragm valves, regulators, and interconnected pipes.

Several cleaning sequences were developed to determine the most efficient cleaning method, including inert gas purging, pressurization, and high vacuum conditions facilitated by pumps. The assessment process involved considering purging time, period, and working pressure as operating variables. The objective was to reduce impurities effectively and minimize cleaning time while ensuring the safety of workers and the working environment.

In addition, the study employed Computational Fluid Dynamics (CFD) simulation to examine the hydrodynamics of residual gas in the system. This simulation enabled the visualization of dead-end zones that are difficult to detect visually. By comparing the residual gas concentration after each cleaning model, the study could determine the effectiveness of the cleaning methods in reducing impurities.

The study's results suggest a cleaning model that can minimize cleaning time and maintain the manufacturing process's efficiency. This is significant for semiconductor manufacturing industries that use specialty gas supply systems, as it provides them with valuable insights into the cleaning procedures that can help maintain impurity levels within an acceptable range.

Overall, this study highlights the importance of efficient cleaning methods for specialty gas supply systems in the semiconductor manufacturing industry. Using CFD simulation and assessing operating variables can aid in developing effective cleaning procedures that can reduce impurities, minimize cleaning time, and maintain manufacturing efficiency. The findings of this study can be useful for industry professionals and researchers alike, as they provide insights into the cleaning models that can help optimize manufacturing processes.

# 85e Amine Blending Optimization for Maximizing $CO_2$ Absorption Capacity in a MEA - Mdea -Water System Using the Thermodynamic Model

*Jinyoung CHA*

This paper reports the During thermodynamic modeling, Kent Ensiberg using equilibrium constants and DM model using activity coefficient electronic NRTL and UNIQUAC models were used. Also newly measured experimental data for $CO_2$ solubility in a blended aqueous solution of Monoethanol amine (MEA) and Methyldiethanolamine(MDEA) at different amine mixing ratios (MEA:MDEA:$H_2O$ = 9:21:70, 15:15:70, and 21:9:70 wt %) and working temperatures (323.15, 343.15, 373.15 and 383.15 K)under CO2 partial pressures of up to 400 kPa. The successive substitution method [1] was used for calculating the mole fractions of all molecules (four molecules) and electrolytes (three cations and four anions) from the equilibrium, along with the material and charge balance equations (11 equations). Using the above mentioned thermodynamic models, the partial pressures of $CO_2$ in the gas phase, mole fractions of all components in the liquid phase, heats of absorption, and cyclic capacities of $CO_2$ according to the absorption/desorption temperature and the blending ratio of MEA/MDEA/$H_2O$ were estimated. We can also predict the ratio that represents the optimal efficiency of the amine. The regeneration energy the sum of sensible heat, latent heat, and heat of reaction, was calculated according to the amine blending ratio using the "shortcut method" [1], the Novel shortcut method suggested by Lee et al. was used to investigate regeneration energy of stripper for process optimization. In the process, besides the cyclic capacity and heat of absorption, the effect of sensible heat and latent heat must be considered simultaneously.Therefore, the most reliable method to determine the amine blending ratio is regeneration energy minimization.

# 85f Techno-Economic and Environmental Assessment of Onsite Green Hydrogen Production By Ammonia Cracking

*Sijan Devkota*

Research Interests

In this study, the green hydrogen is produced from ammonia cracking in a multi catalytic packed bed reactor with an intermediate heating system by using Aspen Plus®. The process results were validated with experimental data for commercially available $Ru/Al_2O_3$ catalyst. The required thermal energy for the cracking is generated from the combustion of about 9% of feed ammonia. In the ammonia combustion, furnace thermal efficiency, fuel-saving, and product yield are analyzed, and optimum values are found to be 59%, 22% and 77%, respectively. The overall thermal efficiency of the developed process is calculated to be 79%. To monitor the $NO_x$ emissions, the parameters such as equivalence ratio, combustion temperature and pressure are analyzed to witness the optimum operating conditions. The blending of hydrogen with fuel ammonia resulted in higher $NO_x$ emissions, which are reduced by recirculation of 30% of the flue gas with air-fuel stream. Two bed TSA and four bed PSA processes are developed in Aspen Adsorption, recovery of 80% and purity of 99.999% are obtained. Economic analysis is performed, including an itemized cost estimation, sensitivity analysis of process and economic parameters, uncertainty analysis and profitability analysis. The process parameters such as furnace temperature, flue gas recirculation, ammonia decomposition temperature, market ammonia supply pressure, ammonia decomposition pressure, hydrogen purification pressure and equivalence ratio, and economic parameters of CAPEX and OPEX are considered. Cost of hydrogen production is found to be highly sensitive on the price of ammonia. Whereas the LCOH is estimated and found to be 6.05 USD/kg of $H_2$. Monte-Carlo simulation method is applied to present LCOH keeping in mind uncertainties in prices of equipment, energy, and raw material. The profitability analysis is conducted by calculating profitability indicators like ROI, PBP, NPV, IRR and PVR for 25-year plant life with 10% discounted rate. In addition, this study reports global warming potential analysis under different energy source scenarios.

## 263a Duplicating Manufacturing Facilities across Different Continents: Case Study

*Anam Ahmed*

Increasing business resilience, production volume and global market share have driven many companies to consider investing in the construction of newer production facilities across different geographic regions. Existing chemical plants within the business unit tend to serve as the design basis for the construction of the new facilities essentially saving time on detailed engineering. However, these design bases need to be modified to take into account the new region's climate impacts, variability in raw materials available, utilities and labor costs and may be subjected to stringent EHS regulations. All these considerations demand proper design modifications. Any deficiencies can have a big impact on reliable production and can truly undermine the economic benefits of the investment.

This goal of this presentation is to share key learnings from unique challenges and operability issues faced when commissioning a new chemical manufacturing facility and discuss the tools and the processes utilized to achieving operation resilience and securing stable production.

## 263c Scale-Down Considerations for Optimization and Troubleshooting of Operating Processes

*Roque Gochez-Campos*

Dow Consumer Solutions Process R&D (DCS-PR&D) leads the design and implementation of new process schemes, as well as the optimization and troubleshooting of existing operations that may deviate from designed behavior.

Scale-down is the idea of taking an existing manufacturing process (>1000 KG scale) and reproducing it in the lab scale (1 – 30 KG). This approach allows PR&D to troubleshoot and test hypotheses in the lab, knowing that the learnings will be applicable in manufacturing. This is essential to optimize reaction times, validate simulations with experimental data, and perform root cause investigation of reactors that are generating out-of-spec material. Scaling down processes, however, is not always straight forward.

In this presentation, a variety of scale-down considerations will be reviewed, ranging from technical aspects such as thermodynamics and fluid mixing, to effective idea generation between R&D and manufacturing, in order to formulate hypotheses and define research deliverables.

# 263d Lithium Borohydride Quench Characterization and Off-Gassing Management

*Emily Cunningham, Shelby McCowen, Jeffrey A. Nye, Rebecca Green, Scott Savage*

Gas generation is a common occurrence in organic synthesis as gaseous side-products are often generated. To ensure such reactions can be carried out at any scale, proper characterization is needed. Accurate balanced equations are necessary to determine the theoretical amount of gas generated, however analytical methods must be used to quantify the rate and total amount of gas released. In this work, excess $LiBH_4$ from a reduction reaction was quenched using acetic acid, resulting in hydrogen off-gassing. Full quenching of the borohydride species was required to ensure safe scale-up and avoid pressurization during downstream processing. An initial large amount of gas and heat release were observed and controlled during lab development. However, a slow residual release was also observed over 48 hours. This residual off-gassing level was so low that it was below the detection threshold of typical lab-scale gas flowmeters.

A process analytical technique was developed that could effectively and reliably quantify the rate and amount of hydrogen gas generated. Raman was used to monitor the degradation of the borohydride species, and thus could be used as a proxy for hydrogen off-gassing. Once we had a reliable method for tracking quench completion, numerous strategies were investigated to mitigate safety issues prior to scale-up. We demonstrated that heating the quench for an extended time resulted in the complete consumption of excess LiBH4 and the subsequent borohydride species. This approach was implemented in two successful batches with no safety incidents related to the quench or hydrogen off-gassing. Further investigation was done to identify alternative quench methods for use in future campaigns, including sacrificial carbonyls and stronger acids.

## 263e Measurement Science and Calorimetry in Reactive Chemicals Evaluations of Laboratory-Scale Processes

*Jessica E. Nichols*

Over the past half century, Dow's Reactive Chemicals Program has supported a strong safety culture and a scientifically rigorous, risk-based review process for chemical hazards in the company's labs and plants. The foundation of Dow's chemical and process hazard analyses is the data generated by the Reactive Chemicals Group, who leverage measurement science, expertise in chemistry and engineering, and an understanding of Dow's processes to support safe operation at all scales. Research activities at the laboratory and pilot scales represent a unique challenge with respect to managing reactivity hazards and preventing incidents. While a smaller scale can reduce the risks and consequences of an incident, research activities often involve new chemicals with unknown hazards, frequent changes to scope in response to new discoveries, or reactions and materials that are not yet well-understood. In addition, time and resources are often limited, particularly when it is not clear whether the research will be successful. Despite these challenges, identifying, evaluating, and addressing reactivity hazards is essential to prevent injuries or equipment damage from uncontrolled or unexpected chemical reactions. To ensure that research and development of new chemistries and processes continue safely, the Reactive Chemicals Group at Dow applies an agile, flexible, safe, and rigorous approach to reactive chemicals hazard identification and evaluation for small-scale processes. In this talk, I will discuss the fundamentals of applying calorimetric data to assess reactivity hazards. While dynamic or isothermal calorimetry may be appropriate in some hazard evaluations, understanding many worst-case scenarios requires other techniques, particularly adiabatic or pseudo-adiabatic calorimetry. I will review the assumptions and technical limitations that must be considered in these experiments and illustrate the appropriate application and interpretation of example data in the evaluation of a prospective experiment. I will also discuss how a Reactive Chemicals Subject-Matter Expert might predict and evaluate the likely outcomes of plausible worst-case scenarios in several laboratory experiments, with a focus on the effect of scale.

# 263f Primary Reformer Revamp and Modernization of Vintage Ammonia Plant

*Muhammad Waqas Quraishi, Arshad Naveed*

Engro Ammonia plant Primary Reformer was commissioned in 1967 at Pascagoula, USA and was relocated to Pakistan in 1992. Over its operational life of more than 50 years, it has undergone various revamps for capacity enhancements. This primary reformer uses Turbine Exhaust Gases (TEG) as primary oxidant coming from gas turbine. During the last few years, plant load was curtailed due to unstable combustion in firebox mainly due to maldistribution of TEG/fuel resulting in structural oscillations. Detail Engineering was done including Computational Fluid Dynamics (CFD) to identify the bottlenecks.

Several modifications were proposed and implemented in 4Q-2022 turnaround including:

- Telescopic design of TEG duct with improved metallurgy
- Individual burner fuel header size upgradation
- Catalyst tubes replacement with upgraded metallurgy (micro alloy)
- Staged fuel burners installation
- Convection coils adequacy evaluation and replacement
- Forced draft (FD) fan installation along with air preheater
- Addition of trip logics as per HAZOP study

After major revamp of Primary Reformer; plant is operating well with best energy efficiency and reliability as envisaged at design stage. This paper entails impact of modifications on plant reliability and efficiency improvement.

## 164a Microfluidic Photodegradation Studies of Quantum Dots

*Hamed Morshedian, Milad Abolhasani*

Colloidal quantum dots (QDs), with their distinctive size-tunable optical properties, have demonstrated great success in a wide variety of applications, such as optoelectronic devices, photocatalysis, and bio-imaging. Photostability of QDs is an important factor for the long-term applicability of QDs in chemical and energy technologies.[1] Conventional flask-based photostability studies of QDs suffer from non-uniform and difficult-to-control experimental conditions and are labor-, time-, and material-intensive. The process-dependent nature of batch photostability studies of QDs has resulted in a knowledge gap in precise mechanistic understanding of photodegradation of QDs.[2] In this work, we present a material-efficient microfluidic platform for intensified photostability studies of colloidal QDs. Utilizing a single-droplet photoflow microreactor, we demonstrate 3.5X faster photodegradation while using a 100X lower amount of QDs than batch techniques.[3] The developed single-droplet photoflow reactor provides in-situ access to the optical properties of QDs during the photodegradation process under a collimated and tunable high-energy photon flux. Utilizing the developed microfluidic platform, we study the mechanism and kinetics of the photodegradation of cadmium selenide (CdSe) QDs. Our *in-situ* intensified photodegradation studies indicate that singlet oxygen generation via triplet energy transfer from colloidal CdSe QDs can trigger photo-oxidation of CdSe QDs. Our systematic studies using the intensified photoflow microreactor unveil that the photo-oxidation phenomenon causes the etching of CdSe QDs at a constant rate linearly proportional to the incident photon flux. Moreover, a fast initial decay in the photoluminescence of colloidal CdSe QDs is observed and attributed to the emergence of surface trap states via a parallel photodegradation pathway. Next, we studied the effect of the average starting size of CdSe QDs on their photostability and found a higher photodegradation rate for CdSe QDs with smaller average starting sizes. Yet, different photodegradation pathways of CdSe QDs have rates with varying sensitivities to the average starting size. This work provides a detailed understanding of the multiplex and intertwined photodegradation phenomena of colloidal CdSe QDs and exemplifies the unique characteristics of microfluidic approaches to intensify the fundamental and applied photodegradation studies of QDs. The mechanistic photodegradation studies of colloidal QDs offered by the developed microfluidic platform are essential for the development of next-gen high-performing QDs for energy and chemical technologies.

References

1. Moon, H., *et al.*, Stability of Quantum Dots, Quantum Dot Films, and Quantum Dot Light-Emitting Diodes for Display Applications. *Adv. Mater.* **31**, 1–14 (2019).
2. Krivenkov, V., *et al.*, Ligand-Mediated Photobrightening and Photodarkening of CdSe/ZnS Quantum Dot Ensembles. *J. Phys. Chem. C* **122**, 15761–15771 (2018).
3. Morshedian, H. & Abolhasani, M. Accelerated Photostability Studies of Colloidal Quantum Dots. *Sol. RRL* **2201119**, (2023).

# 164c Membrane Reactor and Crystallization-Based Process Intensification Strategy for Para-Xylene Recovery

*Nitish Mittal, Jingjun Liu, JR Johnson, Benjamin A. McCool, Prodromos Daoutidis, Michael Tsapatsis*

### Abstract

Para-xylene recovery process is highly energy-intensive due to limited equilibrium conversion of other isomers to p-xylene resulting in a recycle loop of 3-4-fold feed rate, and phase change during isomerization and recovery. Process improvements which can reduce this recycle ratio and/or eliminate the need of phase change operation/vapor phase isomerization can lead to significant energy savings. A novel process intensification concept is proposed for xylene recovery that overcomes the isomerization equilibrium restriction by using a membrane-reactor isomerization unit and a crystallization as a secondary separation process to achieve final p-xylene product.

### Introduction

Para-xylene (p-xylene) is a key petrochemical intermediate and is used almost exclusively for the production of synthetic fibers and poly(ethylene terephthalate) resins, which are precursors for various textiles and packaging materials. However, p-xylene is produced along with other isomers, namely ortho-xylene (o-xylene), meta-xylene (m-xylene) and ethylbenzene, and is recovered using adsorption and/or crystallization processes. The other xylene isomers are then converted into p-xylene via equilibrium isomerization process and this mixed xylene stream is again separated to obtain pure p-xylene generating a loop more commonly known as xylene recovery loop. This overall process is highly energy-intensive due to 1) limited equilibrium conversion of other isomers to p-xylene resulting in a recycle loop of 3-4-fold feed rate, and 2) phase change during isomerization and recovery as isomerization happens in vapor phase while adsorption- and crystallization- based process requires liquid phase. In this study, we describe a novel process which can reduce this recycle ratio and/or eliminate the need of phase change operation/vapor phase isomerization and can lead to significant energy savings.

### Methodology

A novel process intensification concept based on a membrane reactor is proposed to mitigate isomerization equilibrium restriction and using a crystallization as a secondary separation process to achieve final p-xylene product. The membrane reactor overcomes the equilibrium by simultaneously converting m-xylene and o-xylene into p-xylene and selectively permeating p-xylene over m-xylene and o-xylene on the permeate side. This selective permeation also leads to a higher purity of p-xylene enriched stream which enables the use of crystallization as a secondary separation process.

A heat and mass balance model for conventional adsorption-based and the proposed membrane-reactor/crystallization process is simulated in Aspen plus. Energy calculations for crystallization are performed based on the reference [Wilsak, 2001]. The membrane reactor performance is based on the zeolite membranes prepared in Tsapatsis Lab [Liu et al., 2021].

In the conventional process, mixed xylene and heavies stream is fed to xylene column which separates C8 aromatics fraction from the heavies. This C8 aromatics fraction obtained from the overhead is then sent to an adsorption-based p-xylene separation unit. The adsorption unit separates the feed into a p-xylene rich extract and a p-xylene depleted raffinate stream. The p-xylene depleted raffinate is sent to a raffinate column (for separation of desorbent) and is recycled back to xylene loop after reestablishing equilibrium composition via isomerization process. The p-xylene enriched extract is then sent to an extract column (separates p-xylene from the desorbent) and further to a finishing column to remove the residual toluene and other lights, and obtain final p-xylene product.

In the proposed novel process, overhead from xylene column is fed to a membrane reactor. The feed/retentate product is recycled back as feed to reactor. Except for heavies, other components will be either consumed (m-xylene, o-xylene, ethyl-benzene) or permeated (p-xylene, benzene) through the membrane. Thus, a small purge is required for heavies in the retentate. p-Xylene enriched permeate is fed to crystallization unit after distilling benzene in benzene column. Permeate hydrogen is recycled back to the permeate side – a purge or make-up hydrogen can be used to maintain permeate hydrogen based upon the direction of permeation of hydrogen through the membrane-reactor. The filtrate stream from the crystallization is recycled back to membrane reactor. Further, benzene column can be completely heat-integrated with the condenser of the rerun column. Thus, the total energy required constitute the rerun column reboiler, phase change which is associated with membrane-reactor, and crystallization energy.

Accounting for the heat integration, the base case adsorption process results in a total of 245 kW/kTA which is in agreement with 249 kW/kTA based on the utility consumption provided in one of the literatures on adsorption technology [Commissaris, 2003]. The proposed process intensifications including membrane reactor and crystallization suggest that energy requirement can be lowered down to ~135 kW/kTA.

## Conclusion

A novel process intensification concept is proposed for xylene recovery that overcomes the isomerization equilibrium restriction by using a membrane-reactor isomerization unit and a crystallization as a secondary separation process to achieve final p-xylene product. A heat and mass balance model for conventional adsorption-based and the proposed membrane-reactor/crystallization process is simulated in Aspen plus. A significant advantage associated with proposed case is the reduction of the recycle ratio. While an adsorption/reactor base-case configuration recycles 3-4 times the amount of feed, a membrane-reactor/ crystallization recycles only 1.2 times the feed within the membrane and another 1.3 times the feed from the crystallization. This reduction in recycle has potential to lower both the operational and capital costs associated with p-xylene recovery.

## References

Wilsak, R. (2001). Energy efficient process for producing high purity paraxylene. US6565653B2.

Liu, J., McCool, B., Johnson, J. R., Rangnekar, N., Daoutidis, P, Tsapatsis, M., (2021). Mathematical modeling and parameter estimation of MFI membranes for para/ortho-xylene separation. *AIChE Journal.*, *67*, e17232

Commissaris, S. E., (2003). UOP Parex Process. *In Handbook of Petroleum Refining Processes*. 2.47-2.54, McGraw-Hill.

# 164d Multifunctional Reactor for Process Intensification of a Multi-Enzymatic Cascade

*Francesca Meyer, Tim Julian Brandt, Mirko Skiborowski, Andreas Liese, Paul Bubenheim, Thomas Waluga*

## Introduction

Biocatalysis plays a crucial role in the sustainable production of bulk chemicals, specialty chemicals, and high-value products. The use of enzymes, can reduce investment and operational costs and decreases the risk of accidents by enabling milder reaction conditions compared to chemical processes [1]. Enzyme-catalyzed reactions also have high selectivity, increasing yields, and reducing the complexity and costs of downstream processing. However, several challenges still exist, such as high enzyme costs, reactant or product inhibition, and low conversion rates. In order to overcome these, multi-enzymatic cascades can be a sufficient option. They allow the production of tailored products with flexible synthesis routes and can be operated without the separation of the intermediate [2]. In addition to the use of enzymes, process intensification is a key factor in improving biocatalytic processes for sustainable production. As stated by Stankiewicz process intensification "consists in the development of innovative apparatuses and techniques that offer drastic improvements in chemical manufacturing and processing, substantially decreasing equipment [..] and ultimately leading to cheaper, safer, technologics" [3]. In biocatalytic processes, this can be achieved through the use of multifunctional reactors, as e.g. achieved in reactive separations. Multifunctional reactors can be designed to overcome limitations such as enzyme instability or low conversion rates by providing suitable reaction conditions for biocatalysts [4]. Furthermore, a membrane bioreactor can be combined with a packed bed reactor to improve mass transfer and minimize substrate inhibition [4,5].

This contribution investigates the intensification of a biocatalytic process for producing cinnamyl cinnamate, a natural flavor and fragrance. The process employs a multi-enzymatic cascade, as depicted in Figure 1. The first step of the cascade converts cinnamyl aldehyde into cinnamyl alcohol. However, the alcohol dehydrogenase (ADH) used in this reaction requires the costly co-factor NADH. To reduce the use of NADH, a second reaction is introduced for co-factor regeneration using a formate dehydrogenase (FDH). Both reactions take place in the aqueous phase The third reaction in the cascade takes place in the organic phase, where an esterification reaction using an immobilized lipase forms the final product, cinnamyl cinnamate. To ensure optimal reaction conditions, two phases are utilized. The esterification reaction cannot occur in the aqueous phase, as water is produced as a byproduct. The ADH and FDH enzymes are highly sensitive to the organic medium, necessitating the use of an aqueous medium. This reaction system offers an opportunity to produce a natural flavor and fragrance sustainably. However, its technical implementation is challenging as it requires controlling three reactions in two phases. This contribution explores the realization of the process in a miniplant scale and the potential for improving the process using multifunctional reactors.

The shown reaction system was operated in a former project in a miniplant. the previous setup of the process, which comprised a reactor for the two aqueous-phase reactions, an extractive centrifuge for facilitating mass transfer between the phases, and a reactor for the organic-phase reaction [6]. Recent simulation studies identified two primary, independent bottlenecks in this setup: the immobilized enzymes in the aqueous phase and the reflux stream of the organic phase from the fixed bed reactor back into the centrifuge during continuous operation.

## Set-up

The integrated process with modifications for both bottlenecks is shown in Figure 2. When immobilized on a silica carrier, ADH and FDH exhibited only 4% residual activity. Consequently, an alternative to immobilization that keeps the enzymes in place had to be found. An enzyme membrane reactor was suitable for this purpose since it allowed the reaction medium, substrates, and products to pass through while retaining the native enzymes. This can significantly enhance enzyme activity and reaction velocity. The second part for improving the process was to reduce the reflux rate from the organic reactor into the centrifuge during continuous operation. According to Baldea [7], high reflux rates indicated the need for further process integration, such as the integration of the reactor and the extractive centrifuge as a reactive extraction centrifuge. In order to store the organic volume that was not inside the centrifuge, a tank for the organic phase was added. The subsequent sections will describe the implementation of the enzyme membrane reactor, the reactive extraction centrifuge, and the simulations in a miniplant.

Investigation of membrane reactor and reactive extractive centrifuge

Prior to their integration into the process, both the enzyme membrane reactor and the reactive extraction centrifuge were investigated individually. To establish the feasibility of the enzyme membrane reactor, a solvent resistant membrane that offers a high permeate flux, and provides a suitable cut-off between the enzyme and the product size was required. A screening of seven membranes identified an ultrafiltration Pebax® membrane that met these requirements. This membrane was then integrated into an enzyme membrane reactor, which was used to investigate ADH-catalyzed synthesis. Figure 3 illustrates the experimental results of the enzyme membrane reactor, which reveals a sharp increase in alcohol concentration after a dead time of around four hours. The dead time can be explained by the high volume of the membrane plant compared to the volume flow in and out of the plant. The effects of the enzyme membrane reactor were being compared to those of the previous setup, in which immobilized enzymes were used. The experimental results show that the enzyme membrane reactor produced an intermediate alcohol yield that is about 4 times higher than that achieved using the previous setup in the same time span. Adjusting the startup and operation procedures to reduce the system's dead time is expected to increase this effect further.

To study the reactive extraction centrifuge, a modified extractive centrifuge was used as there are no reactive extraction centrifuges commercially available. The reactive zone was integrated into the rotor of the centrifuge, and operation conditions that allow for clear phase separation were determined without conducting a reaction. The volume flow into the reactive extraction centrifuge and the rotational speed were found to be the key factors that influence phase separation, with a rotational speed of 500 rpm and a total volume flow of 500 mL/min being suitable parameters [8]. The whole reactive extraction centrifuge was then studied, with exemplary results presented in Figure 4. These results demonstrate a clear increase in the final product concentration during the operation time, indicating that the novel, internally constructed apparatus is operational.

Simulation of the integrated process

To gain a deeper understanding of the process, it is beneficial to create a mathematical model. In this study, a dynamic model will be implemented in Python with time discretized in one-minute steps. The developed model includes both phases, as well as the concentrations of all substrates, intermediates, and products. To create the model, the reaction kinetics of all three reactions were determined on a laboratory scale, and physical properties and substance-related data, such as the partition coefficient, were used to describe the process [6]. Solubilities were taken into consideration as well. The concentrations, volumes, and temperatures are limited to be zero or higher as a boundary condition. This model can be used to conduct case studies, particularly to determine the initial substrate concentrations and operational conditions, such as the temperature.

Conclusion

To summarize, this contribution explores how the use of multi-enzymatic cascades and process intensification can help reduce the effect of challenges such as high enzyme costs and low conversion rates. Specifically, the intensification of a biocatalytic process for producing cinnamyl cinnamate and the potential for improving the process using multifunctional reactors is investigated. The study proposes the use of enzyme membrane reactors and reactive extraction centrifuges in the process to increase the efficiency of the reaction and to successfully operate the novel apparatus. In addition to that the process is successfully modeled and simulated in order to gain a better understanding of the complex process. This model can be used for further studies to determine bottlenecks and possibilities for further improvement of the process.

References


1. *Green biocatalysis;* Wiley Blackwell: Chichester, West Sussex, 2016, ISBN 9781118822296.
2. Ricca, E.; Brucher, B.; Schrittwieser, J.H. *Multi-Enzymatic Cascade Reactions: Overview and Perspectives*, 2011.
3. Stankiewicz, A. Reactive separations for process intensification: an industrial perspective. *Chemical Engineering and Processing: Process Intensification* **2003**, *42*, 137–144, doi:10.1016/S0255-2701(02)00084-3.
4. *The fabrication of a high performance enzymatic hybrid membrane reactor (EHMR) containing immobilized Candida rugosa lipase (CRL) onto graphene oxide ...*, 2020.
5. Rodríguez-Hernández, L.; Esteban-García, A.L.; Tejero, I. Comparison between a fixed bed hybrid membrane bioreactor and a conventional membrane bioreactor for municipal wastewater treatment: a pilot-scale study. *Bioresour. Technol.* **2014**, *152*, 212–219, doi:10.1016/j.biortech.2013.10.081.
6. Johannsen, J.; Meyer, F.; Engelmann, C.; Liese, A.; Fieg, G.; Bubenheim, P.; Waluga, T. Multi-enzyme cascade reaction in a miniplant two-phase-system : Model validation and mathematical optimization. *AIChE J* **2021**, *67*, doi:10.1002/aic.17158.
7. Baldea, M. From process integration to process intensification. *Computers & Chemical Engineering* **2015**, *81*, 104–114, doi:10.1016/j.compchemeng.2015.03.011.
8. Meyer, F.; Gasimov, N.; Bubenheim, P.; Waluga, T. Concept of an Enzymatic Reactive Extraction Centrifuge. *Processes* **2022**, *10*, 2137, doi:10.3390/pr10102137.

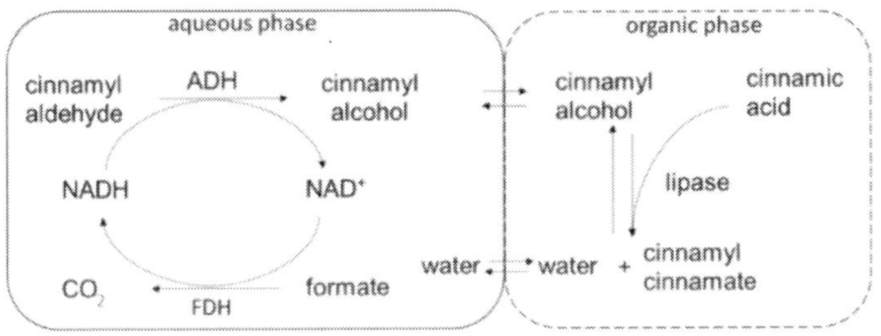


*Figure 3: Multi-enzymatic cascade for the production of cinnamyl cinnamate*

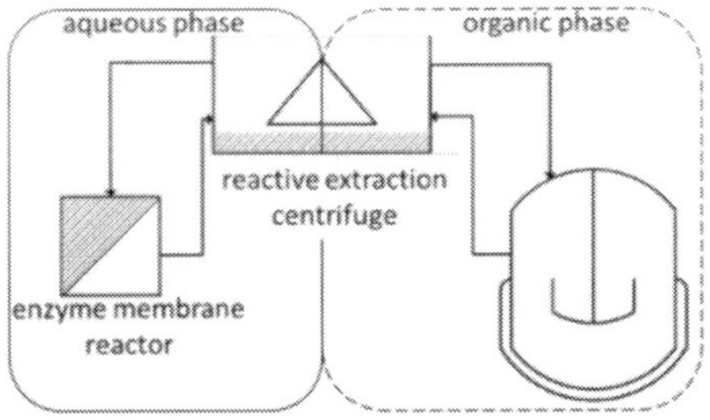


*Figure 2: Set-up of multi-enzymatic cascade in a miniplant scale with improved set-u*

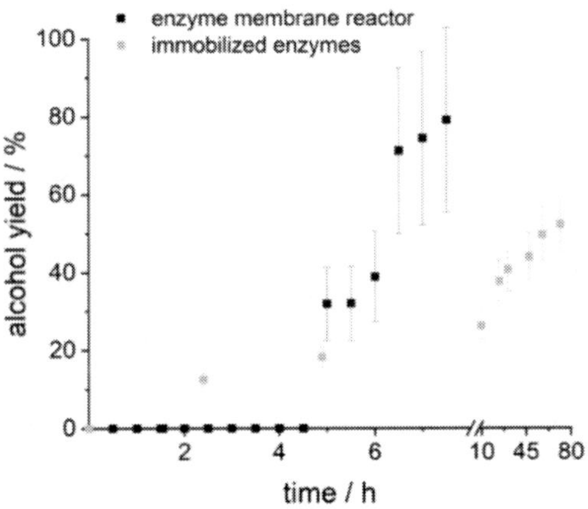


*Figure 1: Experimental results of the enzyme membrane reactor in comparison with experimental data for immobilized enzymes ($V$=0,5 L, $c_{0,NADH}$=3,5 mM, $c_{0,aldehyde}$=3,5 mM, T=25°C)*

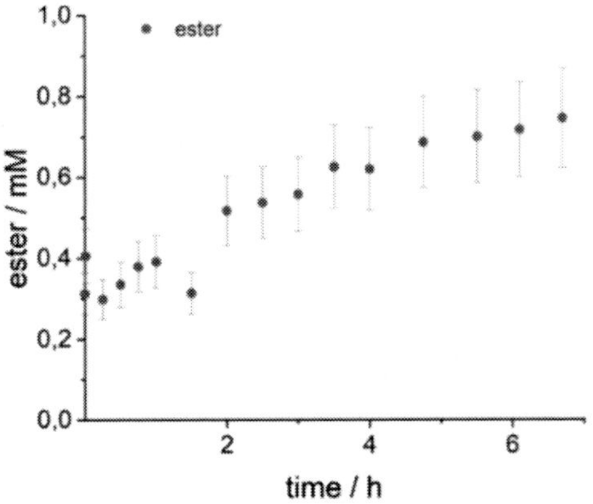


*Figure 4: Experimental results of esterification in reactive extraction centrifuge ($V_{aqueous}$= 0,5 L, $V_{organic}$= 0,4 L, $c_{0,acid}$= 50 mM, $c_{0,alcohol}$= 20 mM)*

# 164e Evaluation of an Electro-Mechanical Piston Reactor for Propane Chemistry

*Aya Abousrafa, Mamoun Al-Rawashdeh, Patrick Linke*

The electrical-driven piston reactor is a new idea that converts electrical energy into mechanical and chemical energy. This reactor has a simple and compact design that can be used to perform chemical reactions. It operates in a similar way to a regular engine by compressing and heating the reactant feed gas or gas-liquid mixture in the reactor cylinder within very short time pulses (milliseconds), creating high cylinder temperatures and pressures that can trigger different chemical reactions and preserve meta-stable desired products. This reactor is based on well-established manufacturing infrastructures since conventional engines exist in various compact sizes and operate at a wide range of speeds, providing flexibility in production capacity. Past research on the piston reactor has mainly focused on using it to partially oxidize natural gas and produce synthesis gas [1]. This reaction is similar to the combustion chemistry used to drive engines, with the only difference being that the exhaust is the desired product.

This work aims to investigate the potential of utilizing the piston reactor driven by electrical energy to carry out propane chemistry. Propane is explored as a feedstock in this work as it produces valuable products such as hydrogen and alkane/alkene products, which are important building blocks for various chemical industries. First, a mathematical piston reactor model is built to predict the performance of the piston reactor running using propane. An experimental setup is built to provide proof of concept for this reactor concept. For this purpose, a single-zone zero-dimensional time-dependent thermodynamic model is developed, incorporating the POLIMI detailed mechanism which includes 484 species and 17790 reactions. The simulation studies showed that high intake temperatures are required to trigger a propane reaction > 950 K. Hence, to initiate the propane reaction within reasonable intake temperatures, reaction triggers such as the addition of oxygen or utilizing a spark plug, are needed

Proof of concept experiments was carried out utilizing a 25-cc piston reactor with a compression ratio of 8:1. Partial oxidation of propane was explored as an initial route. A feed mixture containing air and propane/helium was fed at an equivalence ratio of 1.53, at 1300 RPM, and at atmospheric temperature and pressure. A spark plug was used as a trigger to ignite the fuel mixture within the piston reactor chamber. Analysis of the exhaust revealed that synthesis gas was produced ($H_2$ and CO) with small traces of ethylene and acetylene. Moreover, the exhaust gas temperature due to the exothermicity of propane partial oxidation increased to around 250 °C, indicating the excess heat and power were co-generated.

The developed model was further used to guide experimental work and explore other potential means to carry out the propane chemistry within the piston reactor to produce more value-added chemicals. These studies included investigating the potential of co-feeding $CO_2$, and $H_2O$ with oxygen to operate the reactor within the autothermal region. Figure 1 showcases a case study of one of the simulation results where co-feeding $CO_2$ shifts the product distribution toward the production of methane, and value-added olefins (ethylene, propylene). The simulation data are used to select an optimal operating window and do prove of concept validation experimentally.

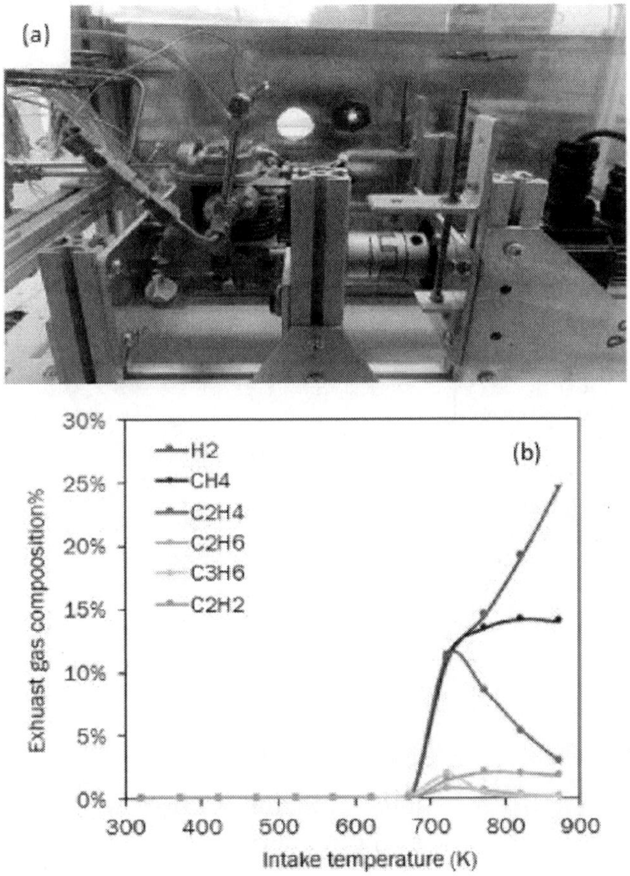


Figure 1: (a) Experimental set-up of motored piston reactor (b) Exhaust gas composition for $C_3H_8/CO_2/O_2$ 1:1.5:0.6 mixture at 1300 RPM, intake pressure = 1 bar, intake temperature = 300-900 K.

# 164f Microflow System for Controlled Synthesis of Ethylene-Vinyl Acetate Copolymers: Continuous Copolymerization and Kinetic Study

*Yuhang He, Yangcheng Lu*

Ethylene-vinyl acetate copolymers (EVA) are important materials with a brand spectrum of application, such as encapsulation material for photovoltaic modules, diesel fuel additives, adhesive and painting materials. In particular, EVA with very high VA content (≥ 80 wt%) is one of the most valuable products among all of the EVA related materials. As it can be used to synthesize ethylene-vinyl alcohol (EVOH), which is one of the most popular gas barrier materials. The Ethylene content in EVA is a key factor that determines its further usage. However, most of the industrial process producing EVA is conducted in batch reactor, which faces the problem of heterogeneous reaction environment and temperature control. As a result, it is difficult to synthesis EVA with well-defined ethylene content in commercial scale.

To solve the problem, a microflow system was proposed to facilitate the synthesis of EVA with controlled ethylene content. The microreactor is reported to excel in mixing, heat removal and residence time control and has been elegantly used to perform reactions under complicated and extreme conditions like multiphase process with improved result. In our current work, we have utilized the ability of micromixer to produce gas bubble to accelerate the dissolution of ethylene in solvent. By doing so, the continuous flow synthesis of EVA by solution free radical copolymerization was realized homogeneously.

The microflow copolymerization process is proved to be better at controlling ethylene content in EVA copolymer by $^{13}C$ NMR spectra, which indicate that EVA obtained in microflow system has less ethylene concentrated sequence than EVA obtained in batch reactor. Ethylene content can be conveniently altered by adjusting the ethylene flow rate or solvent flow rate while not restricted by ethylene pressure. The platform can also be used to investigate reaction mechanism and kinetics efficiently and accurately by operating the microflow system in a stop-flow method. For example, we have developed a simple kinetic of copolymerization and measured the reactivity ratios of ethylene and vinyl acetate copolymerization in tert-butyl alcohol and dimethyl carbonate. The microflow system is demonstrated to be a powerful tool in kinetic and mechanism study and even engineering application of copolymerization processes involving monomers that are gas phase under conventional condition.

## 82a Design Consideration of Multiple Dividing Wall Columns for Renewable Olefins Production Using Economic, Environmental, and Exergy Criteria

*Heehyang Kim, Hosanna Uwitonze, Aejin Lee, Hankwon Lim*

Recently, process intensification has drawn attention as a promising pathway for developing sustainable and cost-effective chemical process systems. Therefore, the technology is being used to improve conventional chemical processes. Likewise, the methanol to olefins (MTO) process which is one of the most important reactions in olefins production also can be addressed to lessen reliance on fossil resources for carbon neutrality. In this work, a renewable olefins production process via synthetic methanol derived from renewable hydrogen and captured carbon dioxide was investigated, and dividing wall column was introduced in the process to reduce cost and energy consumption. The conceptual design and simulation of the process using Aspen Plus and optimization for some parameters were presented. A comprehensive assessment was conducted using economic, exergy, and environmental analyses to evaluate the viability of the renewable olefins production process. Although many papers have conducted the process intensification technology with dividing wall columns, but suggestion of renewable chemical process and comprehensive analysis of the process has hardly been addressed. Therefore, this work has provided some practical and quantitative guidelines for renewable olefins production with dividing wall columns with respect to economic, exergy, and environmental criteria.

## 82b Improved Energy Efficiency and $CO_2$ Emission Reduction in Dimethyl Carbonate Reactive Distillation System through Various Heat Integration Methods

*Minyong Lee, Heecheon Lee, Chaeyeong Seo, Jeongwoo Lee, Jae Lee*

In this study, we propose enhanced energy-efficient design alternatives for the production of dimethyl carbonate (DMC) through the combination of various heat integration methods in reactive and pressure-swing distillation (PSD) columns. The proposed hybrid heat-integrated designs were based on internal and external heat integration between the columns with a vapor recompression heat pump (VRHP). The conventional 3-column DMC production sequence and the hybrid heat integration sequences were compared while maintaining high product purity of 99.5%. Our findings show that the combined sequence of internal and external heat integration led to a significant cost saving of 30.28% by eliminating the low-pressure (LP) column reboiler. The external heat integration combined with VRHP fully eliminated the heat duty of the high-pressure (HP) column condenser and the LP column reboiler, which showed 38.33% energy savings and a 37.5% reduction in $CO_2$ emissions. Furthermore, the two hybrid sequences with external heat integration outperformed the internal heat integration with VRHP in terms of energy, economic efficiency and $CO_2$ emissions due to a significant reduction in the reflux flow rate of the HP column and the reboil flow rate of the LP column.

## 82c Ultrasonically Enhanced Chemical Processes

*Elia Colleoni, Paolo Guida, Deoras Prabhudharwadkar, Chiara Canciani, William Roberts*

Ultrasonically induced cavitation (UIC) can improve chemical processes like wastewater treatment, nanomaterials synthesis, and hydrothermal liquefaction. UIC is a physical phenomenon that entails the formation and subsequent fate of vapor-filled cavities in a liquid medium irradiated by a source of acoustic waves. Bubbles' collapse generates severe local conditions in the continuous medium, accelerating the reactivity of the system.The collapse of a bubble results in the formation of a nano-sized jet and a series of shock waves inducing optimal mixing of heterogeneous mixtures. Another relevant feature of UIC is the physical damage that small particles receive when exposed to nano-sized bubbles' collapse. Despite the beneficial effect of UIC being well known, the implementation at an industrial scale is yet to be achieved. The challenge related to UIC lies in the complexity of the phenomena involved in the process such as homogeneous and heterogeneous nucleation, formation of radicals, and bubbles' collapse. The systems are also generally three-phase systems with a variety of timescales involved.The knowledge gained from either UIC mathematical or experimental activity allowed the design and optimization of UIC reactors for process intensification in three different, but equally relevant, fields: (1) generation of stable water in oil (w/o) emulsions to reduce emissions during combustion. (2) ultrasonically assisted oxidative desulfurization of marine fuels. (3) ultrasonically assisted acid leaching of valuable metals from waste lithium-ions (Li-ion) batteries. A more in-depth discussion of the processes mentioned above is presented in the following paragraphs.

**Water in oil emulsions**: Heavy fuel oils (HFO), a blend of residue from refineries and lighter petroleum cuts, are widely adopted for power generation and marine transportation. It is well known that water-in-HFO emulsions help achieving greener combustion, increasing carbon efficiency and reducing the emissions of pollutants. The quality of the emulsion is mainly a function of three parameters: water content, droplet size distribution, and the stability of the emulsions. UIC induces the formation of nano-scale emulsions, which are stable without the addition of any surfactant. This is possible thanks to two factors. First, asphaltenes in HFO act naturally as a surfactant, disposing at the oil-water interface and creating a rigid layer that stabilized the emulsions. Secondly, emulsions with small droplets are generally more stable than those with large droplets because smaller droplets can resist flocculation, creaming, and coalescence. Thus, the droplet breakup is a key factor determining the droplet size distribution of w/o emulsions. Droplets can be broken by applying energy to overcome the pressure difference across the interface, known as the Laplace pressure. This can be accomplished by using high shearing forces or fluctuating velocities and pressures. Here is where UIC comes into play, allowing tiny droplets and very fine emulsions to form. We successfully applied UIC to generate stable w/o emulsions.

**Ultrasonically assisted oxidative desulfurization:** Oxidative Desulfurization (ODS) is a process that allows removing sulfur-containing molecules from heavy petroleum fractions. The process involves mixing an oxidizing agent with the hydrocarbon mixture, eventually using acid as a catalyst. The oxidizing agent reacts with the sulfur and generates sulfones. The oxidized molecules are then removed via solvent extraction. The solvent, commonly acetonitrile or methanol, removes preferentially sulphones because of their higher polarity. Despite the capability of attacking a wide range of sulfur molecules, the ODS process is more suitable for removing thiophenes, which are generally larger molecules than sulfides. Marine Fuel Oils, generally, present a much larger content by mole of thiophenes, thus making the process appropriate for these types of mixtures. The process operates at low temperatures (330-350 K) and atmospheric pressure. Conversely, hydrodesulfurization, an high temperature and pressure process, is currently adopted to reduce the amount of sulfur in heavy residues. Despite the advantages of the ODS process, various problems affected its performance in the past, precluding any commercialization. Some of those are:

- Low yield
- Difficult sulphones separation
- Extensive usage of oxidizing agent
- Formation of polymers

These limitations were overcome using ultrasonically induced cavitation, resulting in the following beneficial effects:

1. The surface area between droplets of oxidizer and the continuous phase made of oil increases because of the formation of nano-scale emulsions, maximizing the contact between the reactants.
2. The bubbles' formation induces a second reactivity pathway as gas-liquid reactions.
3. The jets induced by the bubbles' collapse destroy the asphaltene aggregates, increasing the probability of exposing sulfur molecules to the bulk. Also, the smaller size of asphaltenes results in better emulsification as they act as a surfactant.
4. Bubbles collapse induces hotspots with consequent radical formation. These radicals enhance the reaction rate.

This allowed us to build a pilot plant to test the ultrasonically enhanced oxidative desulfurization on an semi-industrial scale, with a maximum capacity of 10 tons/day

**Metals leaching from waste Li-ions batteries:** A smooth and rapid green transition to net zero carbon emission requires reducing the use of fossil fuels as energy sources, and creative, evidence-based solutions for more sustainable ways to live, work and travel. Today, one of the most significant gaps in the energy puzzle is the cost-effective and scalable storage of renewable energy to maintain a continuous and reliable energy supply. Lithium-ion (Li-ion) batteries offer the current best-in-practice method for energy storage. Precious metals, such as Nickel, Cobalt, and Manganese, constitute the cathode of Li-ion batteries. However, mining such valuable metals involves high costs and elevated carbon footprint. The limited availability and the controversial geographical location of the known reserves make recycling such metals essential for the near future. Two main industrial processes are currently applied for recycling Li-ion batteries, (1) pyrometallurgical and (2) hydrometallurgical. The pyrometallurgical process is generally characterized by high energy consumption and low recovery efficiency but is easily implementable as a single-step process. On the other hand, hydrometallurgical ensures a higher recovery rate of valuable metals and lower energy consumption. Still, it is more complex as it involves multiple steps and is characterized by using harsh solvents and generates wastewater. According to the current state of the art, the bottleneck of the hydrometallurgical recycling process lies in the leaching step. An in-house developed and optimized ultrasonic reactor allowed to speed up the leaching process significantly, moving its time scale from hours to a few minutes (8 to 10 min). Moreover, the correct application of ultrasound allows moving from solid inorganic acid, currently used in industry, to milder and greener weak organic acid, without mining the performance and the advantages of no UIC-based leaching processes.

The computational tool and the conducted experimental activity allowed us to apply UIC to intensify different relevant industrial processes effectively. Our work focused on the design, optimization, and translation to the industrial scale of UIC reactors. Three test cases where UIC was successfully applied for the process intensification are reported here. The diverse technologies mentioned above take advantage of a particular UIC reactor design that allows modulating the residence time and parameters like time-dependent particle size distribution. The design was guided by a computational analysis performed as a first step of each process scale-up. The computational fluid dynamics (CFD) algorithm was described in previous publications from the group. UIC's potential could be extended further to many different processes, especially those limited by mass transfer phenomena that our ultrasonic technology can easily enhance.

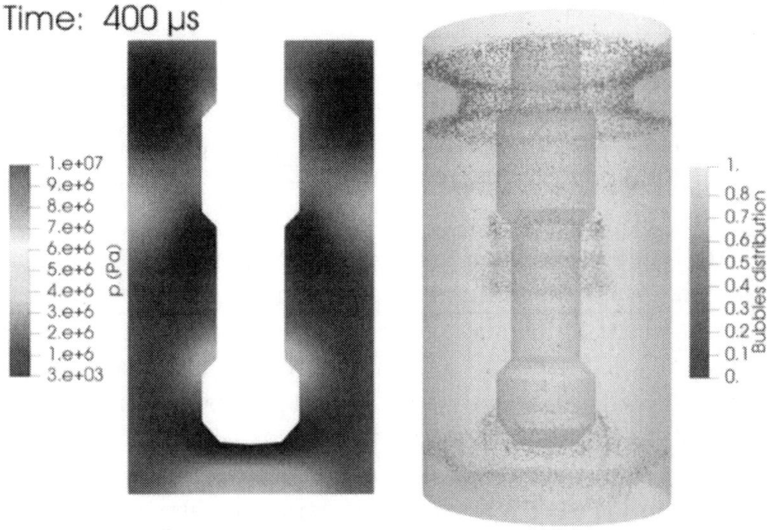


*Pressure field (left) and bubbles field (right) for a given geometry predicted by the model developed in-house.*

## 82e Process Intensification Based on Modular Decomposition to Improve the Sustainability of Chemical Processes

*Arick Castillo Landero, Diana Dominguillo Ramirez, Aburto Jorge, Elias Martinez-Hernandez*

Process intensification through hybrid equipment combining unit operations has the potential to improve sustainability and decarbonize chemical processes by reducing energy and emissions while improving economics. One of the critical steps in intensifying a process is to systematically select which unit operations need to be combined to develop an intensified process. This work presents a novel methodology to guide such intensification using a graph representation and modularity analysis for the decomposition of a process flowsheet such that the process structure itself helps us to identify strongly related process units that have the potential to be combined into hybrid units or other forms of intensification. The methodology aims to identify and analyze intensified alternatives using process graph modularity and the Fire and Explosion Damage Index (FEDI) as criteria to carry out intensification. The first stage consists of developing or reproducing a conventional design through process simulation and obtaining the necessary information to carry out the analysis. In the second stage, safety is analyzed with the FEDI indicator, which is then used as a weighting criterion in the modularity analysis. The intensification of the process is carried out in the third stage, intensifying the process by combining process units that form a process module and generating new alternatives. The fourth stage is used to analyze the alternatives and see if the new designs generate more modules. Process design alternatives are evaluated and compared in terms of energy demand, economic and environmental indicators. Finally, the best alternative is selected. The methodology is illustrated by two case studies, the first involves the separation of an ethanol-butanol-water mixture. The results show that the safest design is the alternative that was developed using this approach and incorporates two columns with a dividing wall and correlates with a high modularity of 0.607. Energy usage is reduced by 25.8% compared to the non-intensified design, which had lower modularity (0.385). An empirically guided design was proposed as Alternative 2 which led to a modularity of 0.533, but only 10% energy savings and no improvement in FEDI. The second case study presents the intensification of a biojet fuel production process. In the intensified process energy utility costs were reduced by 70%, while ROI increased by 20%. In environmental terms, the global warming potential was reduced by 30%. The FEDI value drops from 123 to 105, indicating an inherently safer process. This demonstrates that modularity-driven intensification strengthens integration between process units while improving both the safety and energy efficiency of a process. As such, the approach has wide potential application to guide the intensification of other chemical processes.

# 82f Concept for the Integrated, Quasi-Continuous Production of Crystals – from the First Idea to the Industrial Pilot Plant

*Timo Dobler, Marco Gleiss, Hermann Nirschl*

In almost every industrial sector, companies are faced with constantly changing challenges. In the chemical and pharmaceutical industry, these range from increasing global competition to existing but not exhausted resource and energy saving potentials, growing market volatility, short product life cycles as well as rising quality demands and the desire for individual products in small quantities [1–3]. To meet the challenges successfully and thus ensure the long-term success of the company, innovative solutions are required. In addition to a modular and small-scale design, these include converting batch wise operation into its continuous form, increasing the degree of automation and the integration of different process steps into a single process apparatus [3, 4].

Taking these aspects into account, a novel plant concept for quasi-continuous particle synthesis and separation was developed [5]. A high degree of automation and the combination of different process steps (cooling crystallization, solid-liquid separation and contact drying) in one plant ensure significant energy and resource savings. Furthermore, an intelligent control system in combination with innovative measurement technology enables the targeted adjustment of particle properties and an increase in product quality and quantity.

The basis of the apparatus is a belt filter in which the vacuum trays below the filter medium are replaced by functional modules. While we use temperature control units in the areas of crystallization and drying, filtration segments are installed in the solid-liquid separation section. The individual functional modules can be arranged in any order, which in turn ensures a high degree of plant flexibility and makes it possible to adapt quickly to new material systems or changing customer requirements. The particle formation and separation takes place above the filter medium. To guarantee reproducible, constant process conditions and at the same time to avoid leakage of the suspension during crystallization, the whole operation is performed in zones separated from each other passing through the process chain from left to right. At the end of the belt section, the dried value product is removed from the filter. Figure 1 schematically shows the apparatus concept described.

In addition to developing the apparatus, the focus of the current research was on proof-of-concept and increasing the scale up to an industrial pilot system. For this purpose, tests were initially carried out on a very small laboratory scale. On the basis of the results, it was possible to demonstrate the basic functionality of the concept. Moreover, these experiments served to determine important property functions (e.g. influence of temperature control and residence time on the crystals formed or influence of crystallization on subsequent process steps), to optimize the process, to determine suitable operating parameters and to design an industrial production plant.

Within the scope of the lecture, the complete path from the initial idea to the successful commissioning of the pilot plant will be shown. This includes both the description of the basic apparatus concept and the presentation of results from the laboratory and pilot plant scale. In addition, potential savings resulting from the construction and plant design are discussed and compared with conventional processes.

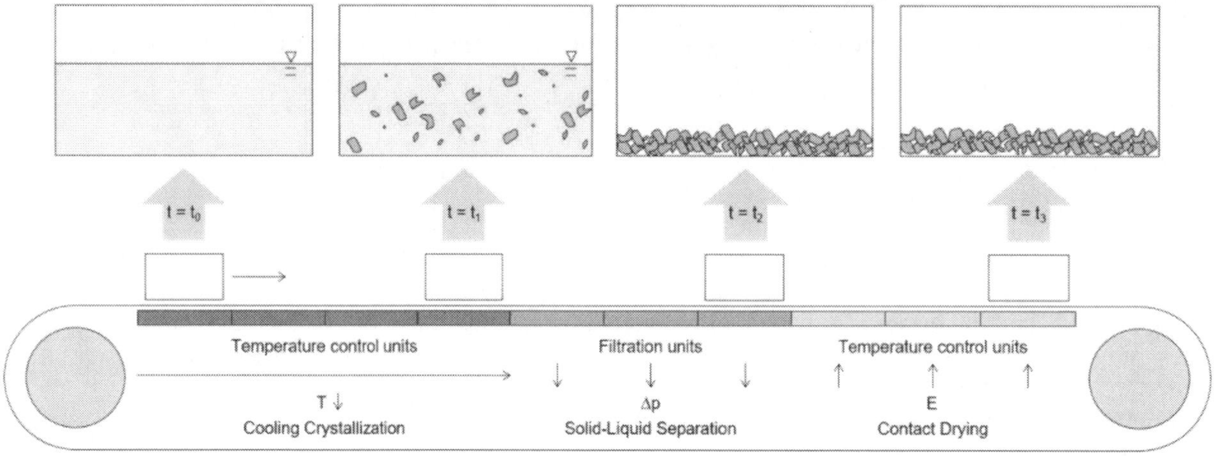


***Figure 1**. Apparatus concept and process procedure. At time $t = t_0$ there is an inoculated and saturated solution. By reducing the temperature $T$, a suspension has formed at $t = t_1$. The suspension is subsequently separated by an applied pressure difference $\Delta p$ and leaves the solid-liquid separation zone at $t = t_2$. Due to the input of thermal energy $E$, only the dried product is left at $t = t_3$.*

## 183a Scale-up of Additively Manufactured Structured Packing Device for $CO_2$ Capture

*Joshua Thompson, Gyoung Gug Jang, Dhruba Jyoti Deka, Amiee Jackson, James E. Parks, Costas Tsouris*

A new structured packing device that utilizes additive manufacturing has recently been designed at Oak Ridge National Laboratory. This device combines external gas-liquid contacting for mass transfer with internal coolant channels for heat transfer that results in process intensification for $CO_2$ capture with amine-based solvents. Previous generations of this device were manufactured and successfully tested at smaller scales with an 8-inch internal diameter column. Employing a new 12-inch-diameter column that was recently set up in our laboratories, we have demonstrated an increase in scale to 12-inches internal diameter. In the current scale-up study of this device, we examined the hydraulic properties of the corrugated geometry, the heat transfer performance under an air-water system, and the $CO_2$ capture performance with and without coolant in its internal cooling channels.

# 183b A Computational Investigation of the Effect of Packing Structural Features on the Performance of Carbon Capture for Solvent-Based Post-Combustion Applications.

*Grigorios Panagakos, Yash Girish Shah*

A major pillar for the decarbonization of point sources emitting $CO_2$ and other impurities is carbon capture, utilization and sequestration (CCUS). CCUS is also vital for the development of the much-needed negative emission technologies. In this work we focus on improving the efficiency of the capture rate of an absorber by designing a device offering in situ cooling. The effective heat management of absorption has been established experimentally as a key pathway for improving overall capture performance, making worthwhile the efforts to comprehend the effect of geometry on conjugate heat transfer coupled to mass and momentum transport.

In our previous work we have successfully incorporated detailed thermodynamics and reaction kinetics into hydrodynamics simulations of the two-phase flow system with an exothermic absorption reaction and generated a few showcases to demonstrate the framework. As a next step we wish now to delineate the effect of the geometry and all related features used to describe it on the metrics that determine the absorber's performance and to promote heat transfer. To that end we have generated python codes to produce a series of fully controlled structures in 2D and 3D that are swiftly imported into our CFD models. For each constructed geometry, we perform CFD simulations by modeling the two immiscible phases as multi-component reacting mixture of MEA–$H_2O$–$CO_2$ solution and flue gas. We numerically solve the transport equations for participating species separately within each phase by explicitly tracking the interface using a mass-conserving volume of fluid method. We characterize the effect of geometrical parameters on the interfacial and wetted areas, liquid holdup, pressure-drop, overall mass and heat transfer coefficients and driving forcers, $CO_2$ absorption rate, and corresponding temperature changes in the solvent for different operating conditions. Post-processing of these data will help guide the development of optimal packing configurations. We use a variety of 2D and 3D Multiphysics simulations to evaluate the accuracy and performance of our approach and to get insights about scaling the technology to higher scales.

## 183c Reduction in $CO_2$-Absorption Column Height and Process Intensification Facilitated By Additively Manufactured Intensified Packing Devices

*Dhruba Jyoti Deka, Gyoung Gug Jang, Amiee Jackson, Joshua Thompson, Costas Tsouris*

Solvent-based $CO_2$ absorption, which is technologically the most matured $CO_2$ capture pathway, still suffers from mid-column temperature rise brought about by the reaction enthalpy that lowers absorption efficiency. This issue can be typically dealt with by an external inter-stage heat-exchanger that cools the solvent; however, this approach increases the overall process footprint, capital, and operating costs. The current study explores a process intensification approach by incorporating inside the column an additively manufactured (3D-printed) intensified packing device that consists of corrugated plates and internal channels; the corrugated plates provide surface area for mass transfer between gas and liquid, while cooling fluid inside the internal channels removes heat from the exothermic reactive system. Not only the intensified device improves $CO_2$ capture efficiency, but it also provides an opportunity to decrease the column height, as each device has the potential to replace multiple commercial packing while keeping the overall performance intact. In the first phase of this project, two different intensified devices with specific surface areas of 250 $m^2/m^3$ and 350 $m^2/m^3$ were designed, manufactured, and tested by positioning, them one at a time, inside a 2.06-m long and 0.2-m diameter column packed with commercial Mellapak 250Y packing. In a second phase, columns with various combinations of intensified devices and commercial packings (ranging from 1 intensified device + 7 commercial packing elements to 4 intensified devices with no commercial packing elements) were tested for their $CO_2$-capture efficiency. A column with four intensified devices with zero commercial packing provides up to 60% reduction in column height. A steady-state kinetic model is also under construction to predict experimental column temperature and $CO_2$ capture performances.

# 183d Microwave Catalytic Gasification of Mixed Plastics and Corn Stover for Clean Hydrogen Production

*Ashraf Abedin, Xinwei Bai, Mark W. Smith, Pranjali Muley*

Of the 353 million metric tons of plastic waste produced worldwide in last two decades, only 9% of plastic waste is recycled, with bulk of plastic waste either landfilled or incinerated. Recycling and upcycling plastic waste are crucial to reduce environmental pollution. Gasification is the thermochemical conversion of solid carbon feedstock at high temperatures with the gasifying agent to generate hydrogen rich syngas. Existing gasification processes are energy intensive and requires high operating temperatures > 1000°C, making the process uncommercial. Microwave heating has attracted interest in a wide range of chemical processes because of its distinct advantages over conventional heating methods, such as quick and selective heating to increase process efficiency. Our research team recently demonstrated the synergistic effects of corn stover and mixed plastics to produce hydrogen rich syngas via microwave heating [1]. To further enhance $H_2$ yield and increase process efficiency, this work reports the addition of magnetite as a bifunctional catalyst to co-gasify plastic and corn stover. Fe catalysts are reported to show high activity in microwave plastic conversion to hydrogen both as a catalyst and microwave energy absorber. Two different sized magnetite was used in this work. A mixture of polyethylene, polystyrene and polypropylene was used as plastic feedstock, while Corn Stover was used as biomass. The catalysts were physically mixed with plastics and biomass at 1:1:1 weight ratio before they were loaded in the reactor to initiate gasification reaction under air flow while varying parameters including temperature and WHSV. The produced gas was directly sent to GC-MS system for analysis and the spent catalysts and char were later collected for deactivation study. The fresh and spent catalysts were analyzed by TGA, BET, Raman and XRD to study catalytic effects for various reaction conditions.

A comparative study was carried out to observe the maximize $H_2$ yield under microwave heating vs conventional heating at similar reaction conditions. The results indicate that under microwave, magnetite produce significantly higher $H_2$ yields compared to conventional reactor under similar reaction conditions. XRD [2] suggested that under microwave, magnetite got reduced (**Figure 1a**) to supply more oxygen radicals in the reaction, which helped to crack tar further down to produce additional $H_2$. Magnetite, being a good microwave absorber, also helped to minimize total power input, which resulted in lower total absorbed energy per gm feedstock, as compared to conventional runs. Decrease in tar and char production was observed in microwave vs conventional run, as the $H_2$ yield increased (**Figure 1b**), suggesting efficient catalytic gasification of the biomass-plastic feedstock.

**Keywords**: Plastic upcycling, gasification, microwave conversion, hydrogen, energy efficiency

**Reference**


1. Ashraf Abedin, Xinwei Bai, Mark Smith, Pranjali Muley, Microwave-assisted co-gasification of mixed plastics and corn stover: A synergistic approach to produce clean hydrogen, Energy Conversion and Management, 280, 2023, 116774, ISSN 0196-8904
2. Gupta, Yashpal & Sharma, Vipin & Kumar B V, Manoj. Issues in Determining Size of Nano-crystalline Particles by X-ray Diffraction. Materials Today: Proceedings. 2, 2015, 3534-3538. 10.1016/j.matpr.2015.07.330.

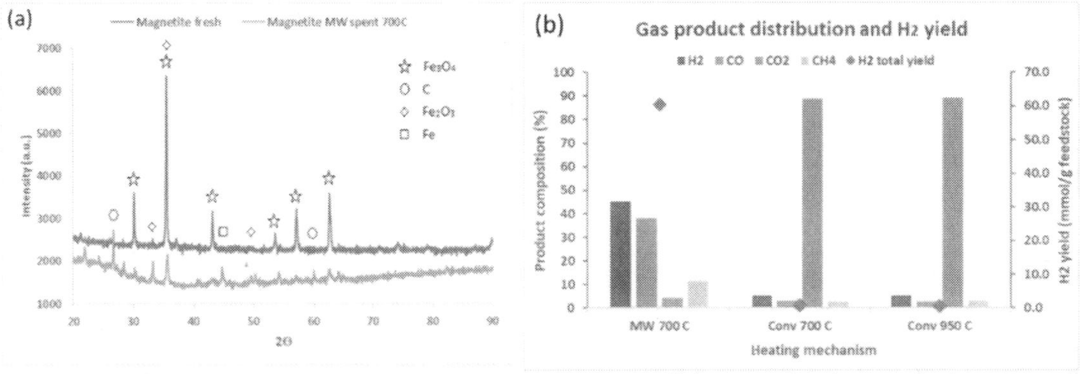


Figure 1: (a) Reduction of $Fe_3O_4$ phase under MW conditions vs fresh sample in XRD; (b) Product composition and $H_2$ yield for MW vs conventional catalytic gasification of plastic and biomass

# 183e Optimization of Catalytic Biomass Gasification Parameters for Cold Flow in a Bubbling Fluidized Bed

*Kshitij Tewari, Changle Jiang, Brandon Robinson, Debangsu Bhattacharyya, Jianli Hu*

This DOE funded project seeks to develop and demonstrate a process intensified two-stage bubbling fluidized bed (BFB) gasifier integrated with a hydrogen separation membrane for the production of exclusively clean hydrogen from biomass. Reactor innovation and process optimization are extremely critical for advancing the biomass gasification technology to produce cleaner syngas with a specific focus on clean hydrogen production. In this work, we designed and build a biomass gasification system (Figure 1) and improved it by optimizing the parameters such as flow to solids ratio, bed length to diameter (L/D) ratio, minimum fluidization velocity, and others. We also evaluated their effects on the performance of gasifier. We studied the impact of gas distributor and catalyst material on the performance of our customized biomass gasifier using a cold flow. The results showed that physical properties, such as size, shape of the gas distributor and size, shape and weight of catalyst material significantly affected the fluidized bed performance. The gas distributor design influenced the flow distribution and residence time, while the catalyst material regulated the reaction kinetics and the quality of the produced gas. The study has provided great insights into the gasification process and reactor design improvements to enhance the efficiency of the gasifier. A study was conducted to investigate the hydrodynamic properties of locally available pinewood as biomass. The study involved both theoretical and experimental methods to determine properties such as particle size diameter, bulk density, and fluidizing velocity. Geldart's distribution diagram was used to categorize the particle size of the biomass residue. The results indicated that the optimal particle size range for good fluidization of pinewood is in between 150-250 micrometers, which corresponds to Geldart's particle size (B). The tests were conducted using a cold flow fluidization bed chamber with varying internal diameters of 19.6 mm, 10 mm, and 8.42 mm, with air serving as the fluidizing medium. The corresponding minimum fluidization velocities were found to be 6.2 litre per minute (LPM), 3.8 LPM, and 2.9 LPM, respectively. The cold flow model will be evaluated in a hot gasifier. In a nutshell, the study showed valuable insights into the reactor design and operation parameters for future research in biomass gasification.

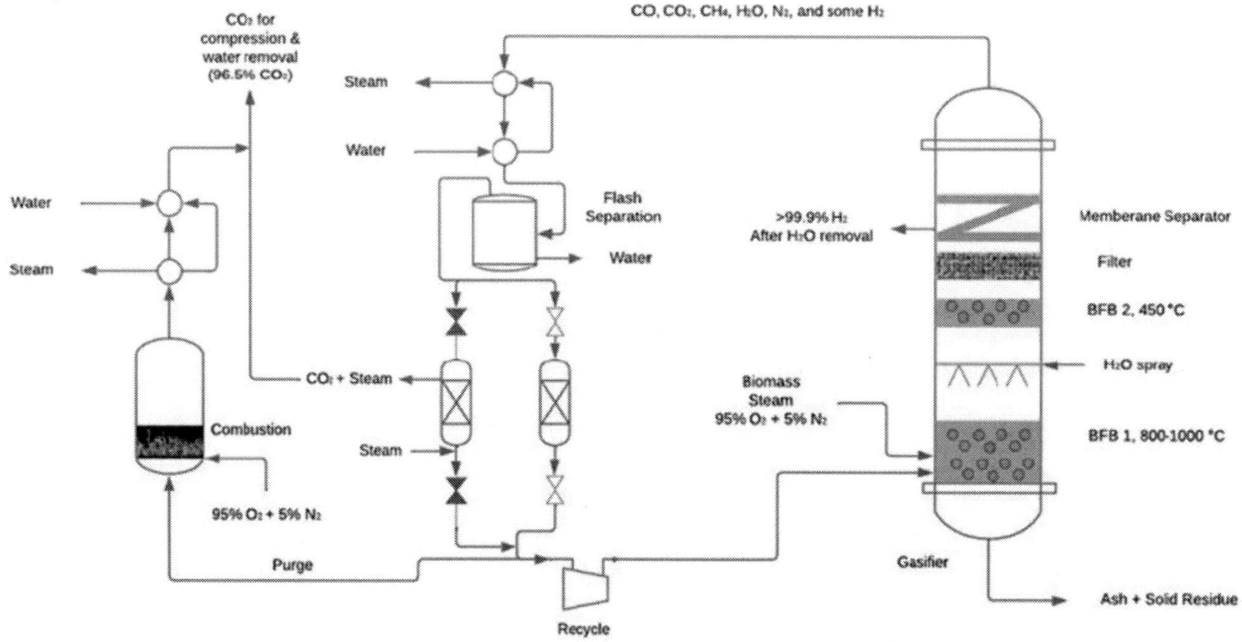

## 183f Lignocellulosic Biomass Valorisation By Technological Coupling of Anaerobic Digestion/Steam Explosion Process

*Oumaima Chaib, Esma Ines Achouri, Nicolas Abatzoglou*

In the alarming context of landfill of organic residues and the generation of greenhouse gases, it is important to develop more sustainable processes to use these residues to produce some renewable energy sources (e.g., biodiesel, ethanol and biogas, etc.). Biomethanation/AD is one the most promising solutions; it consists of a biological degradation of the biomass by living microorganisms in the absence of oxygen. Four steps are categorized in this process: hydrolysis, acidogenesis, acetogenesis, and methanation. the choice of the inoculum depends on the type of biomethanation: Psychrophilic (T=5-25°C), Mesophilic (T=35-40°C) or Thermophilic (T= 55°C). The generated biogas is composed of methane (50 to 75%), carbon dioxide (25 to 45%) as well as some traces of $H_2S$, $H_2$, $N_2$, $NH_3$. The biogas produced can be used for various purposes, e.g., injected into a gas network, used as a source of heat or electricity, used for the synthesis of other value-added chemicals, etc. The technology efficiency depends on the substrate nature; Because of their complex structure (35-50% cellulose, 15-25% hemicellulose and 10-15% lignin), the lignocellulosic residues require a pretreatment step to facilitate the penetration of microorganism, making the cellulose more accessible to the microbial action. Several pre-treatment technologies were tested in the literature: Chemical, physico-chemical and biochemical treatment technologies. However, none of the proposed methods is optimized yet to avoid excessive degradation with production of inhibitor compounds.

The present project aims to improve the management of lignocellulosic waste by developing new technologies whose aim is to increase the carbon conversion contained in this waste, and thus, increase the production of green products such as biomethane. The main idea consists in implantation of a pretreatment step with steam explosion unit preceding the anaerobic fermentation one. This technology is considered as a high temperature thermochemical process (> 200°C), which comprises the destruction and solubilization the biomass allowing a massive hydrolysis of its structure. The yield of biogas production and its composition will be studied according to several conditions of (1) pretreatment of biomass such as: Temperature, catalyst, and residence time in the steam explosion reactor, and (2) anaerobic fermentation conditions such as: Temperature (Psychrophile/Mesophile) Substrate/Inoculum ratio, HRT, OLR.

The lignocellulosic biomass used in this work consists of wood chips (< ¼ inch) from resinous agroforestry residues (spruce) from the Estrie region of Quebec. CANADA. Its composition in sugars, acids, furfural, and 5HMF is measured by an HPLC chromatography. Heavy metals and cations are measured by Inductivity coupled plasma mass spectroscopy ICP-OES. The anaerobic digestion of the pretreated biomass is carried out in a serie of bioreactors operated at mesophilic/psychrophilic conditions. The monitoring of biogas production is done by a manometric method by estimating the volume of biogas produced inside each reactor. The composition of the collected biogas is analyzed by GC (gas chromatography).

The HPLC and ICP-OES results demonstrate the destruction of the biomass and inhibitors production from the degradation of cellulose and hemicellulose. The cellulose degradation releases glucose molecules, which produces 5-HMF. Similarly, hydrolysis of hemicelluloses produces xylose, arabinose, mannose, galactose, rhamnose and acetic acid. In general, at high temperature and high pressure, hexoses (C6) react like glucose to form 5HMF, and pentoses (C5) produce furfural. However, when the humidity is present, the hydration of 5HMF leads to an unstable tricarbonyl intermediate decomposing into levulinic acid and formic acid by hydration. The same samples were used as a substrate for anaerobic fermentation, for 30 days. The important production is showed with using a catalyst at low concentration 0.3%. However, at higher catalyst concentration, the biogas production declines. Without catalyst, the time of SE reaction positively influences the production.

This project is part of the energy transition efforts to respond to the problems of landfill of organic waste. Indeed, the degradation of organic residues generates greenhouse gases including methane and carbon dioxide, responsible for global warming. The scientific relevance of this work is the first attempt aiming to efficiently valorize the residual lignocellulosic biomass by coupling two known processes Steam explosion and Anaerobic digestion SE/AD for sustainable biogas production.

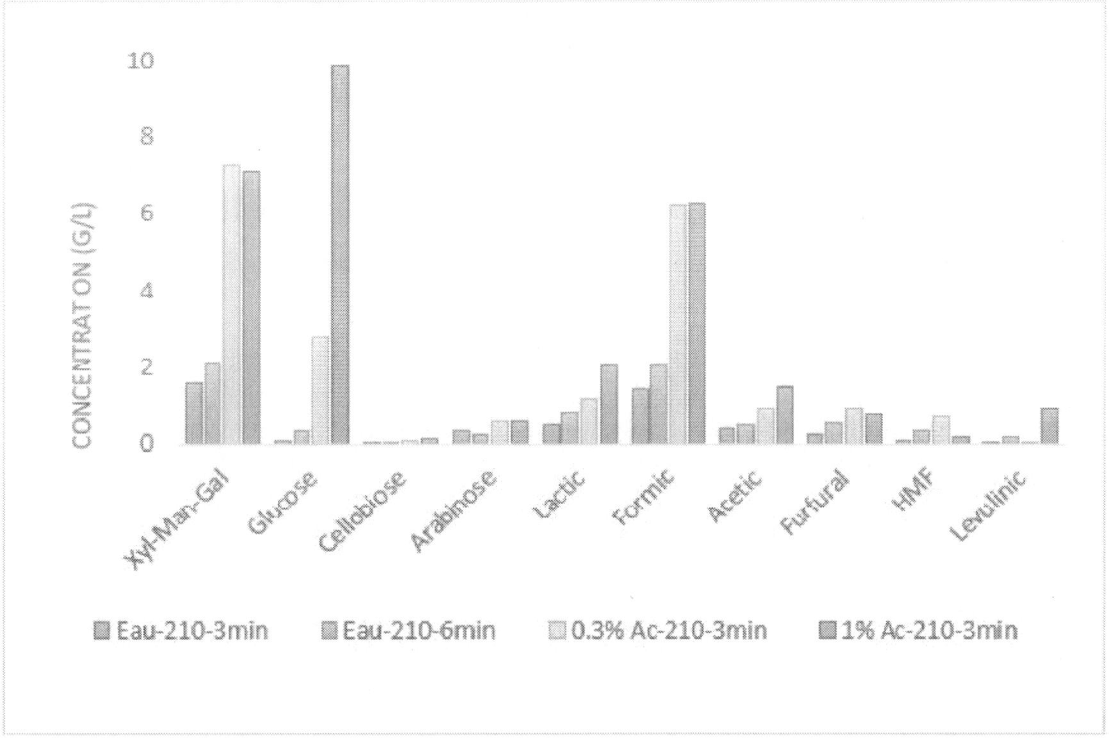


*Figure 1.* Composition of hemicellulose by HPLC as a function of steam explosion conditions

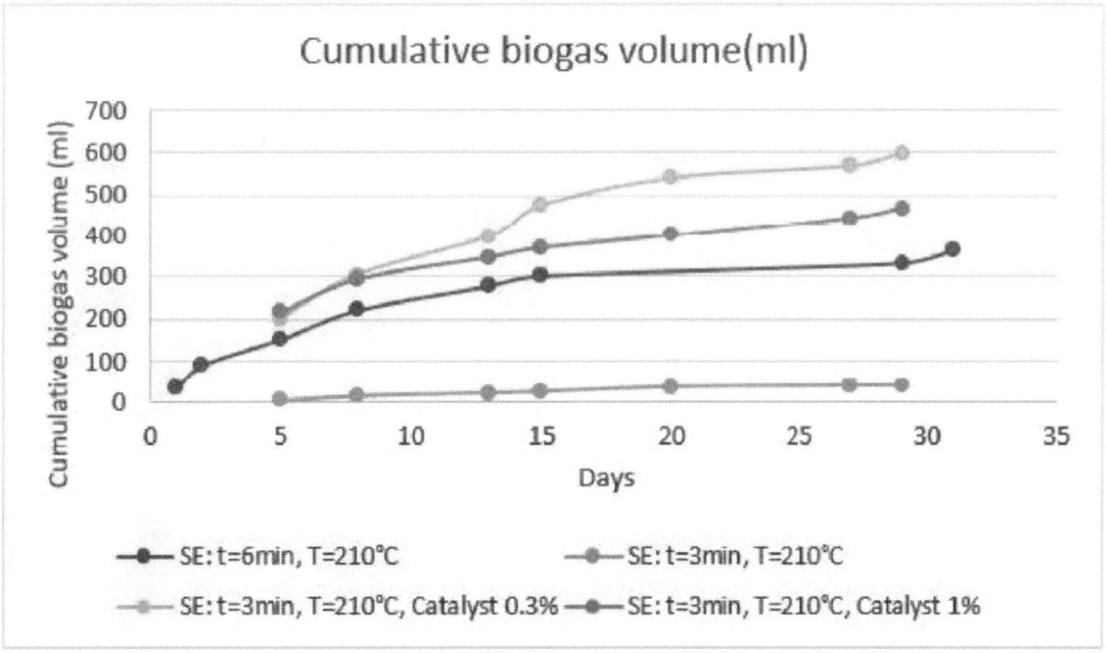


*Figure 2. Cumulative biogas volume as function of SE conditions*

## 104a *Solar-Thermal Dehydrogenation of Propane to Propylene*

*Temidayo Ogunjinmi, Alireza Kianimoqadam, Conor Waldt, David Hibbitts, Justin Lapp, Thomas Schwartz*

This presentation describes an overview of a process concept for significantly reducing the carbon emissions from the conversion of propane to propylene by high temperature catalytic dehydrogenation. The carbon emissions of the existing process are due to the highly endothermic reaction, resulting in reactor temperatures of 600 °C or higher, which is typically accomplished by combustion of refinery gas. We propose a method of providing all thermal energy for the process by solar heat. Pellets of a supported PtSn catalyst are heated by concentrated sunlight to temperatures between 600 and 900 °C. The catalyst pellets are then transferred to a moving-bed, counter-flow reactor where feed pre-heating and the dehydrogenation reaction are driven by heat stored in the pellets. Cool pellets are then recycled to the solar concentrator to be reheated by solar energy.

Key challenges for this decarbonized process include (1) ensuring that catalyst design is suitable for selectivity and low coke formation across a wide range of elevated temperatures to reduce reactor stages, (2) achieving high solar absorptance and stability of catalyst pellets under rapid solar heating, and (3) optimizing throughput in the reactor for effective thermodynamic balance, heat transfer, and reaction time.

Work to-date to address these key challenges has included simulation, synthesis, characterization, and reaction kinetics measurements for a range of silica-supported PtSn catalysts. Data on solar absorption has also been collected for a variety of catalyst formulations and compared with candidate inert solar absorbing particles. A thermodynamics and kinetics analysis has been employed to investigate the proper operating conditions and mass flows for reactors at a variety of scales. Lessons learned to-date and a roadmap for development of this process concept will be presented.

# 104b Microwave-Assisted, Performance-Advantaged Direct Propane Dehydrogenation

*Yeonsu Kwak, Cong Wang, Kavel Chaitanya Ajit, Kewei Yu, Esun Selvam, Reyes Mallada, Jesus Santamaria, Himanshu Goyal, Ignacio Julian, Jose M. Catala-Civera, Weiqing Zheng, Dionisios Vlachos*

Non-oxidative propane dehydrogenation (PDH) produces on-site propylene, a valuable chemical feedstock. Although widely used, its efficiency is limited due to moderate selectivity and significant catalyst deactivation, which makes it operate at lower-than-ideal temperatures. This results in high energy consumption and significant pollution in the chemical industry. We demonstrate that employing microwave (MW) heating with $PtSn/SiO_2$ catalyst pellets in a SiC monolith, acting as a MW susceptor and heat distributor, can enhance the process while maintaining conditions comparable to traditional reactors. Continuous tests show that the MW reactor remains active and stable at 500°C without the co-feeding hydrogen. At higher temperatures, feed concentrations, and with increased space velocity, MW-heated catalysts exhibit remarkable coke resistance, high activity, and exceptional selectivity, unlike conventionally heated reactors that suffer from severe deactivation. Gas-solid temperature gradients are investigated computationally, and nanoscale temperature discrepancies help elucidate the distinct coke formation mechanisms between the heating methods. This strategy highlights the immense potential of electrifying catalytic reactions for eco-friendly chemical production.

## 104c Reactive CFD Modeling, Experimental Validation and Scale-up of Non-Oxidative Ethane Dehydrogenation to Ethylene

*Jeffrey Weissman, Andre DeCarmine, Abhaya Datye, Andrew T. DeLaRiva, Christopher Riley, Alexander Brown, Erik Spoerke*

Ethylene is the largest by volume commodity chemical manufactured world-wide and represents the single largest source of carbon emissions in the chemical industry. Alternatives to ethane or naphtha steam cracking (ESC) to ethylene seek reaction pathways that are lower in energy use and consequently have reduced carbon emissions. One potential pathway under investigation by our team is the non-catalytic non-oxidative dehydrogenation (NODH) of ethane to ethylene, $C_2H_6 \rightarrow C_2H_4 + H_2$. This endothermic reaction is distinct from the related exothermic oxidative dehydrogenation (ODH) of ethane to ethylene ($C_2H_6 + 0.5O_2 \rightarrow C_2H_4 + H_2O$) or the exothermic oxidative coupling of methane (OCM) to ethylene ($2CH_4 + O_2 \rightarrow C_2H_4 + 2H_2O$) as the reactor is simply an empty tube devoid of packings or internal structures, while OCM and ODH require heterogeneous catalysts to proceed. These three reactions share similar gas-phase reaction chemistries; however surface reactions are absent in NODH while important in OCM and ODH, as the latter two depend on the presence of catalysts for reaction initiation [1].

Recently we have demonstrated the effectiveness of NODH for ethylene production from ethane [2]. This non-catalytic endothermic pathway avoids the need to provide superheated steam and complex reactor geometries needed by ESC. Key features of the reaction include use of inert reactor housings, including quartz, and the ability to provide heat of reaction by conventional (external steam, combustion) or non-conventional low-carbon (solar-thermal, concentrating solar power) means. Use of renewable power coupled with less complex downstream separations, due to the absence of steam, can enable up to at least an estimated 40% reduction in carbon emissions associated with ethylene production. Commercial acceptance of our approach depends not only on the reduce energy and carbon impact potentials, but also on throughput and ethylene yield. To this end we used COMSOL Multiphysics computational fluid dynamics software to model ethane conversion and ethylene yields in single reactor tubes capable of producing kg-levels per day ethylene, with a full-scale plant consisting of multiple tubes scaled to metric-tons per day productivity.

An experimental campaign to understand the impacts of diluents (methane, nitrogen), inlet velocities, and reaction temperatures was completed. These measurements were performed in a thermally calibrated quartz reactor tube held in a tubular furnace and at close to atmospheric pressure. Product compositions, including $N_2$, $H_2$, $CH_4$, $C_2H_4$, $C_2H_6$, $C_3H_6$, $C_3H_8$ and C4-C8 species were measured with a micro-gas chromatograph configured and calibrated for refinery gas analysis. No CO or $CO_2$ were measured, verifying that the reaction proceeded in the desired oxygen-free atmosphere. For all tests N2 was used as an internal standard for carbon and hydrogen mass balance closures which were typically from 95-100%. Results were interpreted in terms of ethane conversion, ethylene yield, and selectivities to C3+ species. These results were used to build an irreversible reaction global kinetics model describing the primary features of the results. Experimental first order kinetic parameters (Ea/R and Af) were calculated for the following reactions, written to fit first-order irreversible rate expressions:

- $C_2H_6 \rightarrow C_2H_4 + H_2$
- $0.5C_2H_6 + 0.5H_2 \rightarrow CH_4$
- $C_3H_8 \rightarrow C_3H_6 + H_2$
- $C_2H_4 \rightarrow 0.5C_4H_8$
- $0.5C_2H_4 + 0.5C_4H_8 \rightarrow 0.5C_6H_{12}$

Reactive CFD modeling (RCFD) was then used to optimize ethane conversions and ethylene yield while minimizing C3+ selectivity. Parameters considered including reactor length, diameter, external heat exchange fluid temperature, and total inlet flow rate. Kinetic parameters were refined in an iterative fashion to improve match to experimental conversions and selectivities; good fits between model and data were obtained over a full range of inlet experimental ethane concentrations from 5 to 95% and temperatures from 650 to 850 °C. An example of the RCFD results is presented in the Figure. RCFD identified optimum reactor design parameters in terms of tube diameter, length, and flow rate as well as quartz reactor wall temperature. An optimum 68% single-pass ethylene yield at 95% ethane conversion was obtained, byproducts include 20% selectivity to methane and 8% selectivity to C6's. These results compare favorably to current commercial ethane steam cracking of ~70% single-pass ethylene yield. Our RCFD model, validated against experimental data, shows that NODH is capable of meeting current levels of ESC productivity in terms of product ethylene yields, while significantly reducing the carbon impact of ethylene manufacturing.

***Figure Caption**: Ethylene yields (fraction) predicted by RCFD as functions of inlet flow rate from 1000 (top plane) to 10,000 (bottom plane) cm³/min, as functions of reactor length (x-axis) and diameter (y-axis); for a quartz tube reactor operated at nominally 750 °C, 0.7 meter total length, 0.031 meter diameter.*

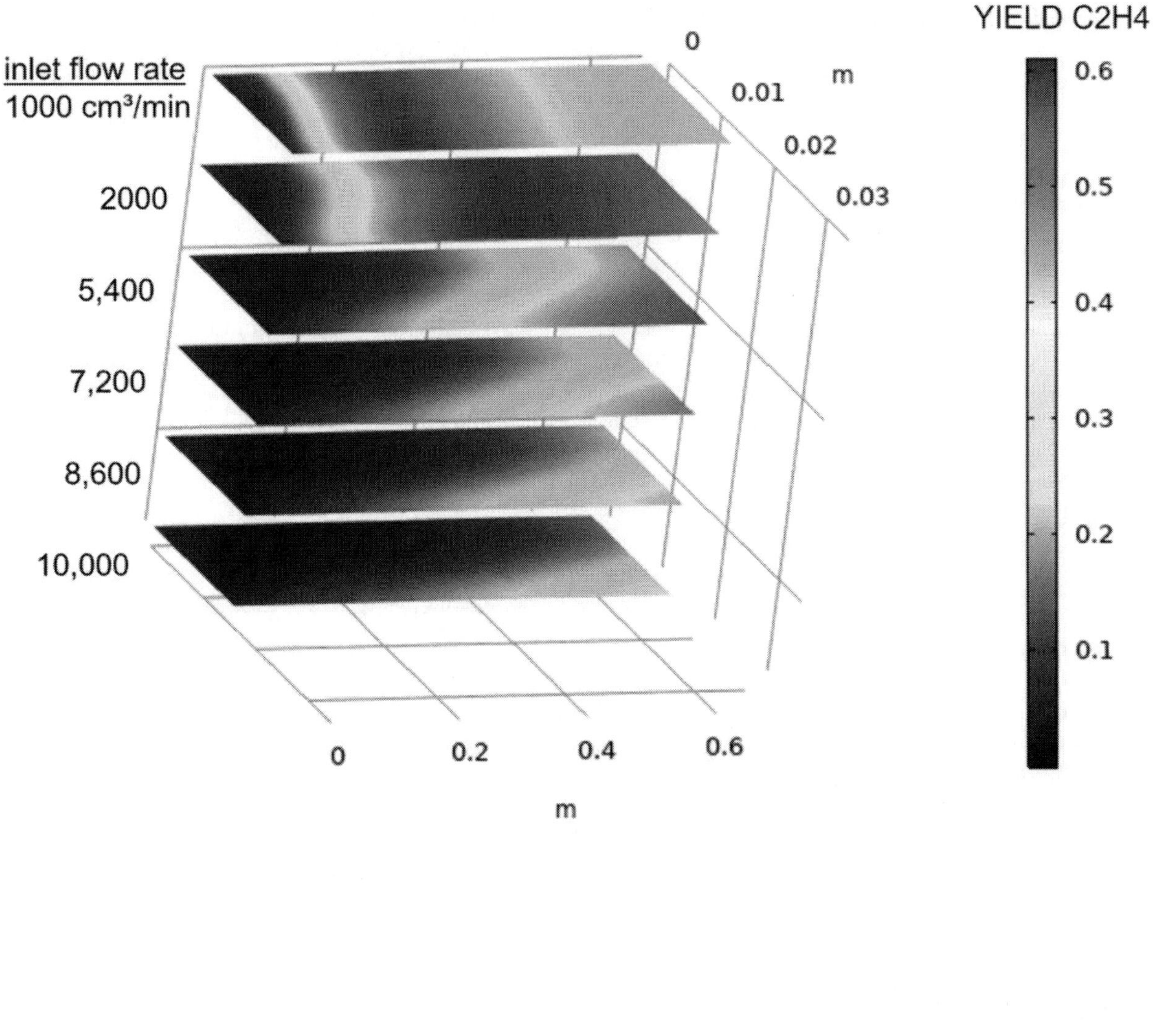

## 104d Carbon Membrane Reactor for Alkane Dehydrogenation

*Lu Liu, Antara Bhowmick, Borja Hernández Blázquez, Marianthi Ierapetritou, Chen Zhang, Dongxia Liu*

Non-oxidative alkane dehydrogenation is an important technology for on-purpose production of alkene and hydrogen ($H_2$), commonly used as building blocks in the chemical industry. The reaction, however, faces three major challenges: the single-pass conversion is limited by thermodynamics, the catalyst deactivates quickly through sintering and/or coking, and the process heat from fossil fuel firing produces copious $CO_2$ emissions. In this talk, we present our recent work on carbon membrane reactor that integrate a carbon molecular sieve membrane and a zeolite supported metal catalyst to achieve high alkane conversion, high catalyst stability and low $CO_2$ emissions. These effects originate from lowering reaction temperature by using catalysts with low threshold temperature and up-shifting reaction equilibrium far beyond thermodynamic limit by $H_2$-permeable carbon membrane. Using propane ($C_3H_8$) dehydrogenation as an example, the membrane reactor enables ~37% $C_3H_8$ single-pass conversion (3-fold higher than equilibrium conversion) and >95% $C_3H_6$ selectivity at 450°C (i.e., ~150°C lower than typical commercial condition), and ~20% reduction in $CO_2$ emissions. Further, we demonstrate that the technology applies to ethane dehydrogenation, establishing a new paradigm toward electrified green and sustainable chemical manufacturing.

## 320a Transportable, Modular Plant for Conversion of Stranded Gas to Low-Carbon Methanol

*Nicholas Schwartz, Paul Yelvington, Edwin Yik, Anthony Dean, Josh Browne, Kyle Merical, Andy Randolph*

Reducing and eliminating methane emissions is critical to mitigating greenhouse gas emissions (GHGs) and responsibly addressing climate change. As the US and other countries rely heavily on fossil fuels, oil and gas wellheads will continue to operate for decades. The majority of the unconventional wellheads in the US are small and remote, which results in the use of flares to burn associated gas (primarily methane) into $CO_2$; however, the conversion efficiency of these flares is limited. Directly converting methane-rich wellgas into liquid methanol helps decarbonize and reduce GHGs while simultaneously producing a valuable product using an otherwise wasted feedstock. Energy efficient, autonomous, and mobile chemical platforms are required to reach the many remote and stranded wellheads in the US and around the world. The approach of a trailer-mounted gas-to-methanol plant addresses the issues with small and isolated wellheads, converting wellgas directly into methanol on site. Modular and transportable units could handle volumes between 65–120 million standard cubic feet per day (Mscf/day), which would yield approximately 800 gallons of stabilized methanol per day. Other sources of GHGs, such as biogas from agricultural operations, landfills, wastewater treatment plants, could also be mitigated with the same gas-to-liquid technology. The M2X Energy process employs an engine-based partial-oxidation reformer technology, tolerant of wellgas contaminants, to convert the feed gas into a hydrogen-rich syngas intermediate that is converted into liquid methanol downstream. Conversion of syngas to methanol is a mature commercial process, which helps mitigate risk and technological development downstream. To develop a mobile and transportable platform, all factors such as throughput, size, weight, energy, and cost must be balanced and addressed. Furthermore, size and weight are coupled with maintenance, accessibility, and manufacturability. Higher wellgas throughput increases the overall size and weight of process equipment, which ultimately limits serviceability and requires special assembly for a finite and constrained mobile footprint. Engineering tradeoffs and size-cost-capacity limitations direct the design for the mobile chemical platform. Optimization of these parameters will lead to a scalable commercial wellgas-to-methanol system that could be deployed at remote wells across the US and worldwide. M2X Energy is currently putting a 65 Mscf/day demonstrator unit in the field to directly convert wellgas into methanol. The demonstrator unit will be used as a platform to scale-up, commercialize, and deploy multiple wellgas-to-methanol systems.

## 320b Integration Options of Natural Gas Pyrolysis in a Methanol Production Process from Biomass: Techno-Economic Environmental Analysis

*Mohammad Ostadi, Guiyan Zang, Leslie Bromberg, Daniel R. Cohn, Emre GenÃ§er*

Low-carbon liquid fuels are needed for transportation and other difficult to electrify sectors. Biomass is a renewable carbon source which holds great promise for generating low-carbon fuel. However, biomass availability is limited, and it is hydrogen-lean; consequently -syngas generated from biomass through gasification is hydrogen deficient. By supplementing external hydrogen, carbon loss from the process can be avoided. Production of hydrogen from natural gas pyrolysis is showing great promise as an alternative to widely known water electrolysis. Natural gas pyrolysis has seen considerable renewed interest in recent years. Because of the endothermic nature of natural gas pyrolysis reaction, the source of the heat required by the process impacts the carbon and thermal efficiencies of the process. Five integration options of natural gas pyrolysis within a thermochemical methanol production process from biomass are analyzed. In addition to heat integration within the process, other integration options examined include natural gas, syngas or hydrogen combustion as well as electrical heating to provide heat for the pyrolyzer. Technoeconomic and environmental analyses of each of these options are performed. From the sustainability point of view, the design where part of the syngas is combusted for heat provision is the best performer while on the economic aspects, it's the least favorite with highest levelized cost. The results confirm the tradeoff between sustainability and economic performance.

## 320c Direct Electrochemical Oxidation of Methane to Methanol on Transition Metal Oxide Electrodes

*Kangze Shen, Ahmed Almunaifi, Carlos Morales-Guio*

Electrochemical partial oxidation of methane to methanol is a promising approach to the transformation of stranded methane resources into a high-value, easy-to-transport fuel or chemical. Transition metal oxides are potential electrocatalysts for this transformation. However, a comprehensive and systematic study of the dependence of methane activation rates and methanol selectivity on catalyst morphology and experimental operating parameters has not been realized. Here, we describe an electrochemical method for the deposition of a family of thin-film transition metal (oxy)hydroxides as catalysts for the partial oxidation of methane. $CoO_x$, $NiO_x$, $MnO_x$, and $CuO_x$ are discovered to be active for the partial oxidation of methane to methanol. Taking $CoO_x$ and $MnO_x$ as prototypical methane partial oxidation electrocatalysts, we systematically study the dependence of activity and methanol selectivity on catalyst film thickness, overpotential, temperature, and electrochemical cell hydrodynamics. Optimal conditions of low catalyst film thickness, intermediate overpotentials, intermediate temperatures, and fast methanol transport are identified to favor methanol selectivity. Through a combination of control experiments and DFT calculations, we show that the oxidized form of the as-deposited (oxy)hydroxide catalyst films are active for the thermal oxidation of methane to methanol even without the application of bias potential, demonstrating that high valence transition metal oxides are intrinsically active for the activation and oxidation of methane to methanol at ambient temperatures. Calculations uncover that electrocatalytic oxidation enables reaching an optimum potential window in which methane activation forming methanol and methanol desorption are both thermodynamically favorable, methanol desorption being favored by competitive adsorption with hydroxide anion.

## 320d Utilization of Nuscale Voygr Integrated Energy System (IES) for the Production of Green Methanol and Synthetic Fuels Via Hydrogenation of $CO_2$ from Direct Air Capture (DAC)

*Francis Tsang, Jose Reyes*

**Title: Utilization of NuScale VOYGR NPM Integrated Energy System (IES) for the production of Green Methanol and subsequent Synthetic Fuels via Hydrogenation of $CO_2$ generated from Direct-Carbon-Capture (DAC) Technology**

**Authors:** Francis Tsang, NuScale Power, LLC, Technical Advisor, USA, Ftsang@nuscalepower.com

José Reyes, NuScale Power, LLC, Chief Technology Officer, USA, jreyes@nuscalepower.com

Luis DePavia, NuScale Power, LLC, Innovation Manager, USA, LDePavia@nuscalepower.com

Methanol, also known as methyl alcohol ($CH_3OH$), is a highly versatile chemical widely used for industrial purposes and prevalent in our everyday lives. It is a base material in acetic acid and formaldehyde, and also increasingly being used in ethylene and propylene. Methanol is one of the most prolific intermediate material for the production of other chemicals and materials. In the chemical industry methanol mainly serves as a raw material in the production of formaldehyde, olefins, acetic acid, MTBE, DME as well as biodiesel. So, renewable methanol is a pre-requisite for making a broad range of chemical products green such as polymer fibers for the textile industry, plastics for packaging, glues, adsorbents/diapers, paints, adhesives, solvents and much more. Besides its use in the chemical, construction and plastics industries, methanol also serves as a fuel or fuel additive. The conventional production method involves a catalytic process using fossil feedstock such as natural gas, coal, or syngas.

Currently, the predominant method of producing hydrogen and $CO_2$ for Methanol, Urea, and Ammonia (Fertilizer) production is the steam-methane reforming process in which natural gas containing Methane ($CH_4$), or syngas, a mixture of $H_2$, CO and $CO_2$, are used both as the feedstock and combustion fuel for process heat. Unfortunately, all these processes result in significant $CO_2$ emissions into the atmosphere.

An environmentally friendly and extremely efficient approach to hydrogenate $CO_2$ would be to use an Integrated Energy System (IES) that includes a NuScale VOYGR nuclear power plant (NPP) for electric power, steam, and thermal production. The process is coupled to a Solid Oxide Electrolysis Cell (SOEC) to generate Carbon Monoxide (CO) gas for the Water-Gas-Shift (WGS) reaction to remove the water produced from $CO_2$ hydrogenation. With the produce water continuously removed from the reaction chamber, the Methanol production can be continued at low pressure and temperature (1 bar, < 400 $^0$C).

The prescribed IES would be coupled to a High-Temperature-Steam-Electrolysis (HTES) system or Solid Oxide Electrolysis Cell (SOEC) Stacks for hydrogen and oxygen production for use in the Carbon Dioxide hydrogenation process. The $CO_2$ required in this process is re-generated within the Carbon Dioxide capturing process of the Direct-Air-Capture (DAC) system. Figure 1 shows the setup and the coupling of the NuScale VOYGR NPP to all the subsystems and components for the production of Green Methanol and the subsequent Synthetic Fuels. The following is a listing of the functions of the subsystems and the intermediate materials that are needed for the entire production process.

- HTSE or SOEC System – Hydrogen and Oxygen Production
- Direct Air Capture (Direct Air Capture) – Carbon Dioxide (CO2) from the Atmosphere
- Reverse Osmosis (RO) water Treatment – Clean water into steam generator to produce steam
- Steam Generator – High quality steam into HTSE for Hydrogen production
- Solid Oxide Electrolysis Cell (SOEC) – convert $CO_2$ to Carbon Monoxide (CO)

**Hydrogen Production -** NuScale performed a study with Idaho National Laboratory (INL) and has teamed up with technology companies to investigate in using NuScale VOYGR NNP for the production of Hydrogen and Oxygen via the HTSE or SOEC technologies (Ref 1 and 2).

**DAC** – NuScale has been involved in the investigation and assessments of capturing Carbon Dioxide from the atmosphere using the liquid sorbent technology. Typical liquid sorbent (Ref 3) is to re-generate the captured CO2 for sequestration. NuScale has been researching the re-use of the captured CO2 for the re-production of useful chemicals.

**RO Water Treatment** – NuScale internal research has demonstrated that RO technology when coupled with NuScale VOYGR NNP generate the smallest carbon footprints (g-$CO_2$ per $m^3$ of clean water) when compare to other energy sources including renewable and hydropower.

**SOEC CO Generation** - The utilization of a SOEC to produce Carbon Monoxide from Carbon Dioxide was patented in March 2013 by a team of Danish researchers from Haldor Topsoe in Lyngby, Demark (Ref 4). It describes a process that $CO_2$ is fed into the fuel side of the SOEC stack with an applied current. On the anode side, purge gas is fed into the stack to enhance the electrolysis process. The output stream from the cathode side of the SOEC stack contains a mixture of CO and $CO_2$. Figure 2 depicts the SOEC setup reproduced from the published patent.

**Methanol production -** At present, most methanol comes from the catalytic conversion of synthesis gas (syngas) that is usually generated by steam reforming of natural gas. The syngas, a mixture of hydrogen, CO, and $CO_2$ is converted into methanol on copper and zinc oxide (Cu/ZnO)-based catalysts at temperatures of 200–300 $^0$C and pressures of 50–100 bar. Steam reforming of natural gas produces copious amount of Carbon Dioxide. Methanol can also be produced by directly hydrogenating pure $CO_2$ with $H_2$ with high selectivity on conventional Cu/ZnO-based catalysts. The reaction rates are much lower than with syngas feeds because of thermodynamic limitations. This synthesis from pure $CO_2$ is also complicated because of the increased water formation. In the absence of CO, water is produced both as the by-product of $CO_2$ hydrogenation. The increased formation of water leads to kinetic inhibition and accelerated deactivation of the Cu/ZnO catalysts. Therefore, the solution is to introduce Carbon Monoxide into the reaction chamber to remove the water via the Water-Gas-Shift (WGS) reaction so that the production of Methanol can be continued. Figure 3 shows the reaction equations and the effects of the WGS reactions.

**Synthetic Fuels and Chemical Production via Methanol** – Methanol as a chemical feedstock for the production of synthetic fuels and chemical products have been studies and widely industrialized globally. Its utilizations are recognized especially for one carbon (C1) based chemical economy. Using steam reforming of natural gas for its production also is recognized as a major drawback due to high greenhouse gas emission. Therefore, Green Methanol production by the NuScale VOYGR NPP methodology can achieve carbon free production for industrial deployments. Figure 4 shows the pathways of Methanol for the production of synthetic fuel and chemical products.

**NuScale Power** Corporation is the industry-leading provider of proprietary and innovative advanced small modular reactor nuclear technology. It delivers scalable, and reliable carbon-free energy. NuScale VOYGR Small Modular Reactor Power Plant system's modularity and flexibility is uniquely qualified to support desalination operations, especially for large-scale production systems. NuScale VOYGR™ SMR plants are powered by the NuScale Power Module™—each a small, safe, pressurized water reactor that can generate 77 megawatts of electricity (MWe) or 250 megawatts thermal (gross), and can be scaled to meet operational needs through an array of flexible configurations up to 924 MWe (12 modules) of output. All the prescribed operations of collecting, processing, concentrating, and ultimately delivering clean water to homes and industries demand a huge quantity of electricity, steam, and thermal energy. The VOYGR SMR plant multi-module configuration layout (4-12 SMRs) is shown in Figure 5.

The NuScale SMR is the only small modular reactor to have its design certified by the U.S. Nuclear Regulatory Commission and published in the Federal Register on 3rd February 2023 (Ref 5). NuScale is well-positioned to serve diverse customers across the world by supplying nuclear energy for electrical generation, district heating, desalination, commercial-scale hydrogen production, and other process heat applications. Figure 6 shows a detailed NuScale VOYGR Plant Site layout.

**References**


- **Idaho National Lab Report, INL/LTD 14-31039, February 2014 (Limited Distribution)**
- **Idaho National Lab Report, INL/EXT-19-55395 Rev.1, September 2019**
- **http://carbonengineering.com/company-profile/**
- **WO2014154253A1;https://patentimages.storage.googleapis.com/49/4b/68/f02e6bfbde9aed/WO2014154253A1.pdf1)WO2014154253A1**
- **https://www.federalregister.gov/documents/2023/01/19/2023-00729/nuscale-small-modular-reactor-design-certification**

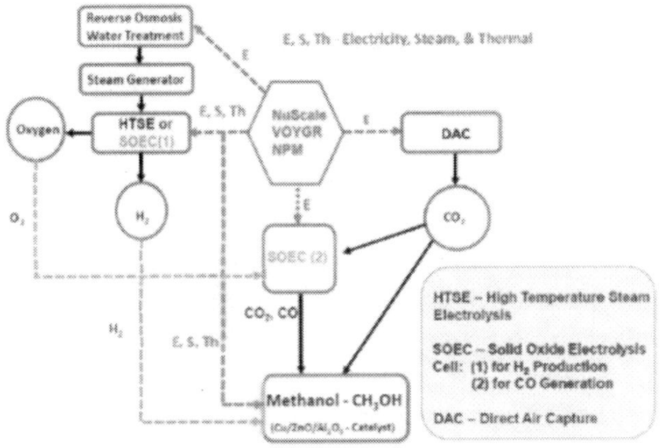


**Figure 1. NuScale VOYGR Plant NPP for Green Methanol Production**

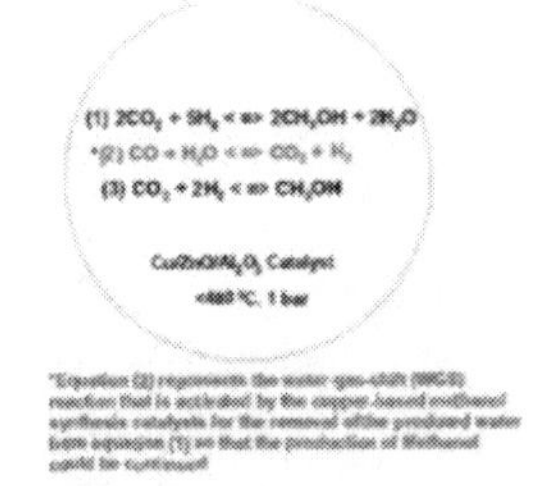


Figure 3. Carbon Monoxide WSG Reaction for Continual Methanol Production

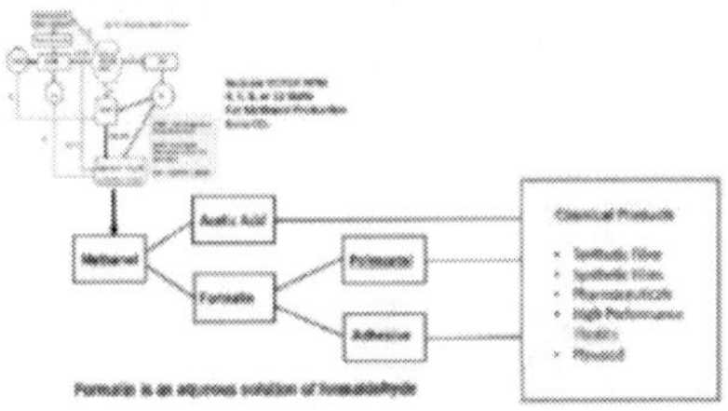


Figure 4. Methanol for the Production of Synthetic Fuels and Industrial Materials

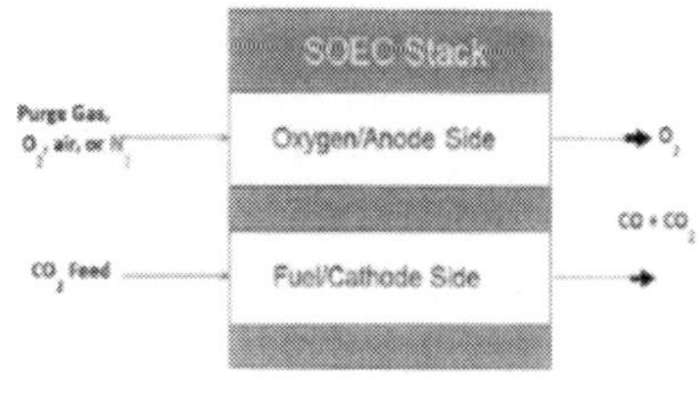


Figure 2. Production of Carbon Monoxide from Carbon Dioxide via SOEC

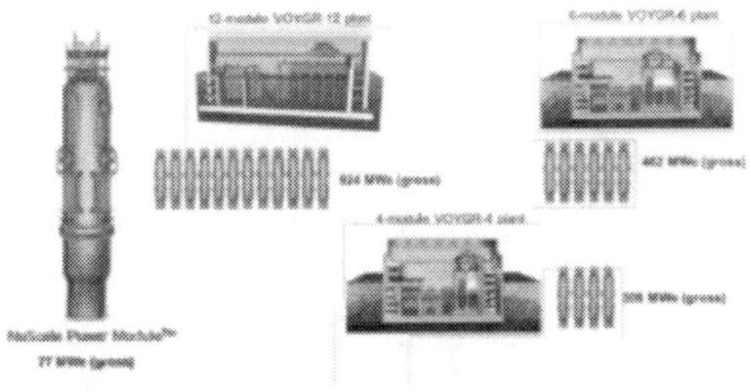

**Figure 5. NuScale VOYGR Small Modular Reactor Configurations**

**Figure 6. Detailed NuScale VOYGR Plant Site Layout**

## 320e Optimal Flexible Power-to-Methanol Processes: With or without Storage?

*Simone Mucci, Dominik Bongartz, Alexander Mitsos*

Methanol is an important platform chemical widely used for the synthesis of several products, e.g., formaldehyde and plastics, and has high potential as an alternative or drop-in fuel [1]. Its production is responsible for about 10 % of the $CO_2$ emissions of the chemical sector [1], which motivates the investigation of sustainable production pathways. A promising alternative is represented by electricity-based methanol (Power-to-Methanol), synthesized from captured carbon dioxide and green hydrogen from electrolysis [2-4]. In this pathway, the high electricity demand for hydrogen production via electrolysis has a high impact on the methanol production cost, thus motivating flexible operation when considering time-variable electricity cost or renewables availability. Additionally, storage technologies, e.g., hydrogen storage, could contribute to addressing these time-variable inputs [5-6]. However, it is unclear if each unit of the Power-to-Methanol plant should be operated flexibly, and which storage technology should be considered to minimize the production cost.

To answer these questions, we implement an algebraic model of a Power-to-Methanol plant composed of the low-temperature water electrolysis, hydrogen compression, and methanol synthesis units, with battery and hydrogen storage in GAMS. We then use the optimizer BARON [7] to solve a combined design and scheduling optimization MINLP problem, which provides the optimal size of the units of the plant when simultaneously considering their optimal operational schedule. Quasi-stationarity at the operating points is assumed due to the relatively high load-change rates of the plant units [2,8] and the hourly time discretization that is considered for the scheduling in the investigated electricity cost and availability scenarios. The combined design and scheduling optimization is performed for a grid-connected and a stand-alone Power-to-Methanol case study by minimizing the annualized cost of methanol.

The optimization results show that flexible operation of both the electrolysis and the methanol synthesis units always reduces the production cost of methanol. Nevertheless, the relative cost reduction between the flexible and non-flexible Power-to-Methanol plants is highly scenario-dependent. Storage, especially hydrogen storage, significantly reduces the methanol production cost in the grid-connected case study when the electricity price is high and highly fluctuating, while the investment cost for storage is not repaid by the savings in the operating costs when the electricity price profile has low fluctuations. In the stand-alone case study, optimally sized storages grant continuous operation of the Power-to-Methanol plant even in case of renewable generation shortage. Also, flexible operation of the whole Power-to-Methanol plant leads to a larger optimal size of the plant units while reducing the need for storage, thus increasing the utilization of the electricity generated by the renewable power plant and the amount of produced methanol.

### Acknowledgments

The authors gratefully acknowledge the financial support by the German Federal Ministry of Education and Research (BMBF) within the H2Giga project DERIEL (grant number 03HY122D).

### References

[1] International Renewable Energy Agency (IRENA) and Methanol Institute, "Innovation Outlook: Renewable Methanol", International Renewable Energy Agency, Abu Dhabi, 2021, ISBN: 978-92-

9260-320-5

[2] V. Dieterich, A. Buttler, A. Hanel, H. Spliethoff, and S. Fendt, "Power-to-liquid via synthesis of methanol, DME or Fischer–Tropsch-fuels: a review", *Energy Environmental Science*, vol. 13, no. 10, pp. 3207–3252, 2020

[3] B. Anicic, P. Trop, and D. Goricanec, "Comparison between two methods of methanol production from carbon dioxide", *Energy*, vol. 77, pp. 279–289, 2014

[4] É. S. Van-Dal, and C. Bouallou, "Design and simulation of a methanol production plant from $CO_2$ hydrogenation", *Journal of Cleaner Production*, vol. 57, pp. 38–45, 2013

[5] J. Gorre, F. Ruoss, H. Karjunen, J. Schaffert, and T. Tynjälä, "Cost benefits of optimizing hydrogen storage and methanation capacities for Power-to-Gas plants in dynamic operation", *Applied Energy*, vol. 257, p. 113967, 2020

[6] C. Chen and A. Yang, "Power-to-Methanol: The role of process flexibility in the integration of variable renewable energy into chemical production", *Energy Conversion and Management*, vol. 228, no. 8, p. 113673, 2021

[7] BARON Solver, https://www.minlp.com/baron-solver (last accessed 29.03.2023)

[8] M. Kopp, D. Coleman, C. Stiller, K. Scheffer, J. Aichinger, and B. Scheppat, "Energiepark Mainz: Technical and economic analysis of the worldwide largest Power-to-Gas plant with PEM electrolysis", *International Journal of Hydrogen Energy*, vol. 42, no. 19, pp. 13311–13320, 2017

# 320f Decarbonization of Natural Gas to Methanol Conversion Process Using Microwave-Assisted Dry Reforming

*Omar Almaraz, Srinivas Palanki, Jianli Hu*

Methanol is a colorless, flammable liquid that is used as a solvent, fuel, and feedstock in the production of a wide variety of chemicals and materials. Methanol can be used as an alternative fuel for internal combustion engines and can be blended with gasoline to reduce emissions. It is also used as a fuel in fuel cells, which convert chemical energy directly into electrical energy. It is a key raw material to produce formaldehyde, acetic acid, methyl tert-butyl ether (MTBE), and other chemicals. These chemicals are used in the production of plastics, paints, textiles, and other products. Methanol is used as a solvent in many industrial applications, including the production of resins, varnishes, and dyes. It is also used as an antifreeze in automotive applications because of its low freezing point and high boiling point. Methanol can be used as a means of energy storage, either by using it as a fuel or by converting it to hydrogen for use in fuel cells. The global methanol market size was valued at \$28.78 billion in 2020 and is projected to reach \$41.91 billion by 2026[1].

Methanol has traditionally been produced from natural gas by first converting methane to syn gas and then converting syn gas to methanol[2]. However, this is a very energy intensive process and produces a significant amount of the greenhouse gas carbon dioxide. Hence, there is motivation to look for alternative routes to the manufacture of methanol. In this research, a decarbonization strategy is developed for Natgasoline's existing process plant that currently converts natural gas to methanol. To achieve this goal, the following specific objectives are pursued: (1) Investigate the impact of replacing the conventional syngas reactor with a microwave reactor for dry reforming of methane[3], (2) Develop a heat integration approach for the entire process to minimize utility costs and improve process efficiency[4], (3) Conduct a techno-economic analysis to assess the cost-effectiveness of implementing the microwave reactor compared to the conventional syngas reactor, (4) Investigate the environmental impact of the new decarbonization strategy and quantify the reduction in carbon emissions achieved.

In this research a novel microwave reactor is used for simulating the dry reforming process to convert methane to methanol. The objective is to produce 14,200 lbmol/h of methanol, which is the current production rate of methanol at Natgasoline LLC in Beaumont, Texas (USA) using the traditional steam reforming process. The feed to the process is natural gas that goes through a desulfurization step to remove sulfur and the resulting methane stream is sent to a novel microwave syngas reactor, which operates at a temperature of 800 °C and pressure of 1 bar to produce a mixture of carbon monoxide and hydrogen. Additional hydrogen is added to this syngas stream and sent to a methanol reactor train where methanol is produced. These reactors operate at a temperature range of 220-255°C and pressure of 76 bar. The reactor outlet stream is sent to a distillation train where the product methanol is separated from methane, carbon dioxide, hydrogen, and other products. The carbon dioxide is recycled back to the microwave reactor.

This process was simulated in ASPEN Plus. The thermodynamic property method used was RKSoave for the process to convert methane to syn gas and NRTL for the process to convert syn gas to methanol. A base case simulation with no heat integration shows that the feed stream of 5,115 lbmol/h of methane results in a final product stream that consists of 14,200 lbmol/h of methanol that has a purity of 99.99%. This novel process was compared with traditional steam reforming, and it is observed that the microwave process utilizes only 36% of the methane as compared to the traditional steam reforming process. Furthermore, there is significant decarbonization potential as this process *utilizes* 9,401 lbmol/h or carbon dioxide whereas the steam reforming process *produces* 575 lbmol/h of carbon dioxide. Furthermore, there is significant reduction in hot utility and cold utility usage in the novel microwave process. Heat integration tools are utilized to reduce the hot utility and cold utility usage in this integrated plant that leads to optimal operation. In particular, the use of a heat exchanger network (HEN) is considered in this research. The optimization of HEN is based on the objective function of minimizing the total utility costs and the minimum temperature approach adopted is 1°C. To optimize the heat exchanger units, heat exchange area and loads on each utility a mixed-integer-linear-programming (MILP) approach is used. Several feasible solutions are obtained using the MILP heat integration tool that is built-in ASPEN Energy Analyzer. Simulations are conducted in this heat-integrated plant to study the effect of changing feed conditions on the product quality and quantity. The simulation results from this study demonstrate the significant potential of utilizing a microwave-assisted reactor for dry reforming of methane. There was a noteworthy decrease in the amount of methane utilized, indicating a more efficient and sustainable process. Furthermore, the microwave process demonstrates the potential for substantial decarbonization of the process, contributing to the overall reduction of carbon emissions. The simulation results also revealed the existence of substantial potential for heat integration in the microwave-assisted process. Overall, the findings of this study highlight the importance of utilizing innovative technologies such as microwave-assisted reactors and process integration techniques in the chemical industry to enhance process efficiency and sustainability.

## Acknowledgment

This study was supported by the United States Department of Energy.

## References

[1] Methanol Market by Feedstock (Natural Gas, Coal, Biomass), Derivative (Formaldehyde, Acetic Acid), End-use Industry (Construction, Automotive, Electronics, Solvents, Packaging), and Region - Global Forecast to 2028, Markets and Markets. (2023). https://www.marketresearch.com/MarketsandMarkets-v3719/Methanol-Feedstock-Natural-Gas-Coal-30408866/

[2] G. Aimiuwu, E. Osagie and O. Omoregbe, "Process simulation for the production of methanol via CO2 reforming of methane route," Chemical Product and Process Modeling, 0049 (2020)

[3] T. T. P. Pham, K. S. Ro, L. Chen, D. Mahajan, T. J. Siang, U. P. M. Ashik, J. Hayashi, D. P. Minh, D. N. Vo, "Microwave-assisted dry reforming of methane for syngas production: a review," Environmental Chemistry Letters, 18(6), 1987-2019 (2020)

[4] M. E. Haque, N. Tripathi, and S. Palanki," Development of an Integrated Process Plant for the Conversion of Shale Gas to Propylene Glycol," Industrial & Engineering Chemistry Research, 60 (1), 399-41 (2021)

# 143a Safety and Procedural Enhancements of Flushing Operations in Petroleum Pipelines

*Swapana Jerpoth, Barnabas Gao, David Theuma, Sean Curtis, Michael Fracchiolla, Robert Hesketh, C. Stewart Slater, Mariano J. Savelski, Kirti Yenkie*

The petroleum lubricants industry faces the unique challenge of product cross-contamination during sequential production and packaging operations. These industries manufacture and package a variety of unique lube oil compositions throughout the year (Parkash 2009). A multiproduct pipeline system is used for sequentially packaging over 2000 batches of final products depending upon the customer demands (Zlokarnik 2001). Therefore, between each change-over operation, there is a need to clear these lines and ensure product purity (Baptista et al. 2000). Most of the straight-pipeline sections can be cleared out using tools such as a piping inspection gauges (PIG), but there are more complex pipe networks with variable diameters and auxiliary equipment that cannot be cleaned using traditional methods (Tiratsoo 2013). These industries must use pure oil to clear these unpiggable sections of the residual oil in the lines to prepare for the next product. This process of flushing generates mixed oils that has low economic value and functionality (Li et al. 2023; Camussi, Cerdá, and Cafaro 2021). The goal of this work is to minimize the amount of mixed oil produced in each changeover by optimizing the flushing process (Baptista et al. 2000; Bamoumen et al. 2022).

Our project focuses on applying creative solutions to reduce the amount of mixed oil produced in each flush. Our research has revealed that minor procedural changes can make significant improvements in the flushing operations, with little to no additional capital costs for the stakeholders. The effect of temperature, efficacy of pressurized air blowing, and safety concerns all are contributing factors when designing an improved procedure. The finalized procedure incorporates air-blowing through the pipelines to reduce the amount of residual oil left in the lines and therefore reducing the amount of new oil required. These procedural changes were tested on a newly developed benchtop pilot plant; less flush oil was required using the air-blowing enhancements. This new procedure then underwent a scale-up analysis for a larger diameter section of pipe where the improved procedure still proved viable, safe, and effective for an industrial-scale packaging plant. These newly developed procedural enhancements can revolutionize pipeline flushing operations and benefit petroleum industries worldwide.

Originally, the study conducted had a focus on identifying the economic benefits of optimizing the operations of these packaging plants. However, the safe operation of these lines is a significant concern as it directly links to 'good manufacturing practices' and workplace safety guidelines implemented by regulatory agencies such as OSHA (Occupational Safety and Health Administration). Some of the hazards identified include aerosolization of harmful lubricant products, improper procedural enforcement, and inconsistent product characteristics.

The identification of the hazards associated with this process is only the first step. Addressing the hazards in a way that does not interfere with the day-to-day operations of the facility is equally important. In order to study the effect of these hazards on operator safety, a series of benchtop experiments were performed. These experiments simulated the physical environment of the operator, including spacing out process controls to mimic the accessibility in a real plant. This involved placing sensors and control valves that mimic the real plant and its constraints in regards to space and accessibility. These procedural changes were also tested to determine their operational safety.

The changes in the benchtop scale system were recorded and the flushing experiments were conducted using the standard operating procedures. Furthermore, the detailed economics and process safety indicators were calculated and compared against the base case. Thishighlighted both the economic and process safety benefits of the procedural enhancements. Developing easy to understand standard operating procedural outlines, combined with operator training is a critical factor in implementing changes that improve process safety. To test the efficacy of the solutions presented, the team conducted in person operator training over a three-month period at the partnered packaging plant. By systematically identifying process hazards, swift actions were taken to ensure the safety of the operators, while also providing an economic benefit to the production plant.

**References:**


Bamoumen, Meryem, Selwa Elfirdoussi, Libo Ren, and Nikolay Tchernev. 2022. "Continuous Time and Volume Batch Formulation for the Multiproduct Pipeline Network Scheduling Problem." *IFAC-PapersOnLine* 55 (10): 2282–87. https://doi.org/10.1016/j.ifacol.2022.10.048.

Baptista, Renan Martins, Felipe Bastos de Freitas Rachid, and José Henrique Carneiro de Araujo. 2000. "Estimating Mixing Volumes Between Batches in Multiproduct Pipelines." In *Volume 2: Integrity and Corrosion; Offshore Issues; Pipeline Automation and Measurement; Rotating Equipment*, V002T08A008. Calgary, Alberta, Canada: American Society of Mechanical Engineers. https://doi.org/10.1115/IPC2000-247.

Camussi, Nélida B., Jaime Cerdá, and Diego C. Cafaro. 2021. "'Mathematical Formulations for the Optimal Sequencing and Lot Sizing in Multiproduct Synchronous Assembly Lines.'" *Computers & Industrial Engineering* 152 (February): 107006. https://doi.org/10.1016/j.cie.2020.107006.

Li, Yansong, Dong Han, Baoying Wang, and Jianfei Sun. 2023. "An Industrial Scale Simulation Method for Predicting the Contamination Evolution in Long-Distance Multiproduct Pipelines." *Journal of Pipeline Systems Engineering and Practice* 14 (1): 04022054. https://doi.org/10.1061/(ASCE)PS.1949-1204.0000696.

Parkash, Surinder. 2009. *Petroleum Fuels Manufacturing Handbook: Including Specialty Products and Sustainable Manufacturing Techniques*. 1 edition. New York: McGraw-Hill Education.

Tiratsoo, John. 2013. *Pipeline Pigging & Integrity Technology, 4th Ed.* Beaconsfield: Clarion Technical Publishers.

Zlokarnik, Marko. 2001. *Stirring: Theory and Practice*. Wiley. https://doi.org/10.1002/9783527612703.

## 143b An Experimental Study of Multiple-Phase, High-Temperature, High-Pressure Reactions between Hydrocarbons and Sulfur

*Peter Koronaios, Hseen Baled, Goetz Veser, Ryan Weber, Nico Proust, Josh Harntraft, Glenn Cormack, Robert Enick*

Sulfurized hydrocarbon compounds have many applications as anti-wear additives and are typically produced in industry in batch reactors. In order to allow manufacturers to make sulfurized hydrocarbons in small continuous tubular reactors rather than large batch reactors, the reaction time must be substantially reduced by changing the reaction conditions.

A novel, high-temperature, high-pressure, three-vessel system was developed and used to run the reactions at temperatures up to 300 C and pressures to 10,000 psi. The hydrocarbon, which was retained liquid under pressure, was held in one variable-volume vessel at the desired temperature. The sulfur was held in a windowed, variable-volume, stirred reaction vessel and was heated to the desired temperature. A third, quench vessel, at room temperature, was connected to the reaction vessel. Once the temperature and pressure were stable, the hydrocarbon was moved to the reaction vessel to begin the reaction. Once the desired reaction time was reached, the contents of the reactor were moved into the cold quench vessel to rapidly stop the reaction. The reaction products were analyzed and the conversion was determined. The results and findings of this experimental study indicate that a batch to continuous transition is possible. A summary of the results will be presented.

## 143c Challenges Faced in Upgrade and Expansion Projects of Existing Chemical Facilities.

*Anam Ahmed*

The ever-changing landscape of chemical industry especially post-COVID economic times, have created opportunities for production of diverse chemical products. Chemical industries are shifting their focus to creating new value-added product streams with a wide range of applications to improve product portfolio, facilitate long-term growth and enhance business viability. In order to realize these economics benefits, companies have to make strategic investments to expand production facilities. However, process design studies for such expansion projects pose a level of uncertainty – they are risky with unique challenges.

Such projects require more intense feasibility studies that pose such challenges which include, but are not limited to:

- Examination of existing infrastructure
- Regulatory compliance
- Complex PHA's due to integration with existing processes
- Updates to existing data analytics models
- Close collaboration between the different business units

Such activities naturally tend to require more time and resources. This presentation will cover many such challenges, share best practices and tools that help minimize expenditure and prevent pitfalls during the process development and implementation stages of expansion projects.

# 143d Understanding the Mixing Mechanisms to Enable Sustainable Manufacturing of Bioinspired Nanomaterials

*Richard Hodgkinson, Siddharth Patwardhan, Mohamed Ashar Yousuf*

Nanomaterials have a vast potential in applications such as food, coatings, cosmetics, textile, transport, healthcare, and electronics, with a multibillion market. While many new nanomaterials are being discovered, their manufacturing is unsustainable and therefore their potential is not realised. Emulating natural designs, we have developed sustainable approaches to produce high value nanomaterials. To reach the market, it is crucial to develop a strategy to scale-up their manufacturing. With the example of bioinspired silica (BIS), we report on understanding the underpinning mixing mechanisms and develop a methodology for scaling-up a wide range of nanomaterial synthesis. Herein, we take a two-pronged approach: on one hand we investigated reactive mixing during BIS synthesis in a stirred tank reactor and on the other hand, we are investigating mixing using a multi-inlet vortex mixer (MIVM). Combining both approaches, we are generating mixing knowledge for BIS synthesis over a range of production scales.

To this end, a non-invasive colorimetric method coupled with novel image analysis was developed to monitor the reaction and the mixing (*ACS Eng. Au* 2023, **3**, 17–27). Specifically, the technique involved image analysis of colour (pH) change by using a custom-written algorithm to produce a detailed pH map. The degree of mixing and mixing time were determined from this analysis for different impeller speeds and feed injection locations. Cross validation of the mean pH of selected frames with measurements using a pH calibration demonstrated the reliability of the image processing technique. The results suggest that the bioinspired silica formation is controlled by meso- and, to a lesser extent, micro-mixing. Based on the new data from this investigation, a mixing time correlation is developed as function of the Reynolds number. Further, we were able to correlate the effects of mixing conditions on the reaction and the product.

While these results provide valuable insights on mixing occurring during BIS synthesis, the stirred tank geometry can be limiting in terms of energy dissipation and hence mapping wider range is not possible. In order to address this issue, further work is performed using multi-inlet vortex mixers (MIVM). MVIM provide a wide range of flow conditions ($Re$ = 10-10,000) and energy dissipation ($\epsilon$ = $10^{-1}$ – $10^{3}$ W/kg), covering the lab-scales as well as plant scales, hence they are very useful in this research. Further, MIVM can accommodate unequal flow of reactants and can be used for mixing condition requiring flexibility in the inlet stream content and flow rate ratios.

The design also enables the separation of reactive components prior to mixing which is not possible with alternate confined impinging jet mixing geometries. The MVIM was characterised using the residence time distribution (RTD) and the Villermaux-Dushman reaction approaches. Next, the flow properties, the BIS formation process and the particle properties were investigated using a systematic experimental approach. These results from stirred tank reactor and MVIM (V-D and BIS synthesis) and their comparisons will be presented in order to develop mixing knowledge and scale-up rules.

# 143e Increasing Resource Efficiency in Process Engineering on Planetary Roller Extruders By Use of Scaling Algorithms, CFD-Simulation and Prediction Models

*Mario Radwan, Sulamith Christiane Frerich*

The reduction of raw material consumption is a contribution to global environmental protection and can be achieved in two ways. On the one hand, by recycling products and recovering the materials used. On the other hand, by optimizing production processes. The potential of the latter is illustrated by the production volume of fossil based virgin polymers from 2021, which was 352 million tons per year [Pla22]. The vast majority of polymers are produced using extrusion processes. For each product manufactured, various preliminary tests and extensive experimental studies were carried out to determine the optimum process conditions to meet ideal product properties and highest quality standards. Thus, reducing the amount of material used in process development represents a mighty tool to reduce the overall consumption of polymers. This work contributes one way to optimize operating conditions of polymer processing by extrusion using planetary roller extruders PREs.

PREs feature a modular set-up. Each module consists of a roller cylinder with one central spindle and multiple planetary spindles (see figure attached). All components feature a 45° helical gearing which provides radial mixing and axial material conveying at low shear rates, due to the rotating spindles. PREs offer a multitude of configuration options. This includes, for example, a variable number of modules, which allows a precise adaption of the plant to the requirements of the production process. In addition, the geometrical dimension of the plant can also be varied. This is relevant for the process design as the plant size influences possible mass flows. PREs are available from laboratory scale (throughput: 0.5 kg/h - 10 kg/h) to large production scale (throughput: more than 8 t/h). Due to its small dimensions and thus less mass that needs to be maintained at a certain temperature, the laboratory scale extruder has a significantly lower energy consumption. Although this is quite advantageous in terms of resource efficiency in process development, operating parameters of PREs on laboratory scale cannot be transferred directly to PREs on pilot scale. Since the plant dimensions are so different, the maximum possible throughput for example differs considerably, too.

To avoid this problem and to determine comparable process settings, scaling algorithms are needed. These algorithms have been newly developed for PREs and will be presented as part of this contribution. Using them, experimental results from laboratory scale can be transferred to production scale. Thus, process development becomes more sustainable, as experiments can be carried out in small scale with low energy demand and less material consumption.

The functionality of these algorithms was verified with a series of physical experiments processing polymer granules, both carried out on laboratory and pilot scale. The identical experimental setup (see attached figure) consists of PREs with two modules [Rad23]. In the first module, the granules are melted because of external heat supply and internal friction between the rotating plant components. The second module contains a variable spindle configuration in order to be able to analyze the influence of the plant setup on the operating parameters. Various sensors for measuring pressure, temperature and residence time allowed a detailed analysis of the extrusion process. In addition to the plant setup, an analysis of the process settings is also carried out. The throughput (3.75 kg/h – 8.75 kg/h in laboratory scale, 30 kg/h – 70 kg/h in pilot scale), the set temperatures of external temperature control units (185 °C – 220 °C) and the conveying speed (60 rpm – 200 rpm) are varied according to a statistical design of experiments. The polymers processed as received are high and low density poly(ethene) (PE-HD 53090, Total Polymers and PE-LD MP20, Versalis) and poly(propene) (PP-H 9081, Total Polymers), showing different viscosity values.

Although the sensors mentioned above determine all measurable process conditions inside the PRE within physical experiments, non-measurable variables such as the shear rate cannot be determined. Nevertheless, an analysis of these properties is relevant as well, as it allows a more detailed investigation of the influence of extrusion on the material quality. Polymer damage for example may occur as a result of extensive shear rates. Due to their high relevance, the non-measurable variables and the flow profiles within the roller cylinder are calculated within 3D-CFD-simulations.

The results from about 2,000 individual physical experiments and digital simulations are collected in a database. This allows further analyses of statistical significance of the influence of input variables (such as throughput or conveying speed) on output variables (such as pressure or temperature).

In addition, artificial intelligence is used to investigate the scalability of PREs. With the large amount of data recorded, process parameter prediction models were trained using MATLAB-based Machine Learning algorithms.

From the selection of available Machine Learning algorithms, Gaussian Processes and Support Vector Machines (including the respective submodels) have proven to be most suitable for describing pressure, temperature and residence time of PREs. This can be described using the example of process pressure and the Machine Learning quality parameters RMSE and $R^2$. Support Vector Machines predict the processing pressure over a test dataset with a $RMSE_{Pressure,SVM}$ of 0.604 ± 0.125 MPa. The pendant of the Gaussian Processes is a $RMSE_{Pressure,GP}$ of 0.389 ± 0.016 MPa. The coefficient of determination is higher for the Gaussian Processes ($R^2_{Pressure,GP}$ = 0.64 ± 0.03) compared to the Support Vector Machines ($R^2_{Pressure,SVM}$ = 0.10 ± 0.38). To increase the prediction accuracy, especially for Support Vector Machines, a preliminary Machine Learning-based approach was further enhanced by developing a hybrid approach of Machine Learning algorithms including physical models. Thus, the new prediction model architecture consists of parallel structures using both physical and prediction models. If physical models are used, usually they show several simplifications that yield distinct deviations between calculated results and measured values. In the hybrid approach, the prediction model is no longer trained to predict process variables, but the deviation between the physical model and the measured values. The overall result, which is the prediction of the process variables, results from the addition of the physical model and the predicted deviation value. This complex algorithm improves the accuracy of the prediction results in comparison with real measured values. This is particularly evident in the coefficients of determination: the hybrid Gaussian Processes accuracy increases to $R^2_{Pressure,hybridGP}$ = 0.86 ± 0.05, and the accuracy of Support Vector Machines increases to $R^2_{Pressure,hybridSVM}$ = 0.57 ± 0.18.

The consolidation of the developed hybrid Machine Learning prediction models and the 3D-CFD-simulations have proven to be suitable for reducing resource-intensive physical experiments. Thus, physical experiments can be carried out on a laboratory scale. By using the scaling algorithms that have been newly developed, they can be transferred to pilot scale or industrial scale. Thus, this approach with its innovative tools helps to transform process engineering towards more sustainability by digitalization.

[Pla22] PlasticsEurope AISBL, *Plastics – the Facts 2022*, Brussels **2022**.

[Rad23] Radwan, M.; Frerich, S.; *Chem. Eng. Technol.* **2023**, 46 (00), 1-8. DOI: 10.1002/ceat.202200523.

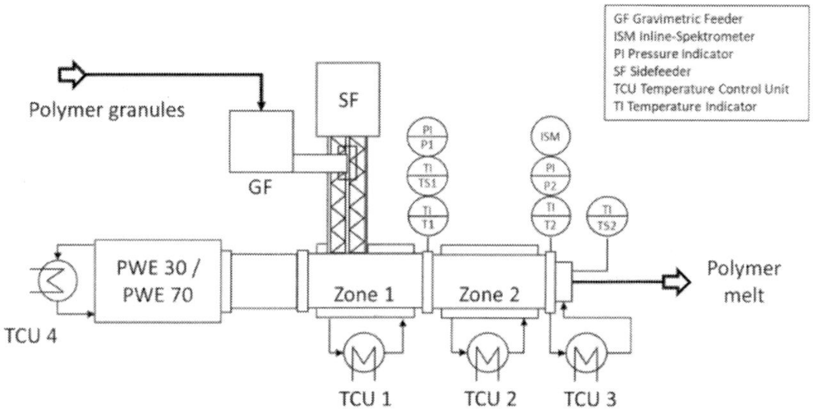
GF Gravimetric Feeder
ISM Inline-Spektrometer
PI Pressure Indicator
SF Sidefeeder
TCU Temperature Control Unit
TI Temperature Indicator
Polymer granules
SF
GF
PI
P1
TI
TS1
TI
T1
ISM
PI
P2
TI
T2
TI
TS2
PWE 30 /
PWE 70
TCU 4
Zone 1
Zone 2
Polymer
melt
TCU 1
TCU 2
TCU 3

[Rad23]

# DSMC Simulation of Axial and Radial Spreading of the Feed Gas in a High Speed Rotating Cylinder


**Sahadev Pradhan**

**Chemical Technology Division, Bhabha Atomic Research Centre, Mumbai-400 085, India.**


## ABSTRACT:


The main focus of this work is to characterize the axial and radial spreading of the feed gas introduced into the high speed rotating cylinder at different axial and radial locations for wall pressure in the range 20 to 100 m-bar using two dimensional Direct Simulation Monte Carlo (DSMC) simulations [1, 2, 3, 4, 5, 6, 7, 8, 9, 10, 11, 12]. The feed gas is accelerated through collisions with the surrounding rotating gas molecules, and there is a slow-down of the rotating gas molecules near the feed injection point, and the slow-down ($V_{\theta,\ solid\ body\ rotation} - V_\theta$) decreases with the increase of axial distance away from the feed point in a very complex manner, with initial slow reduction, and then rapid decrease to zero value. An important finding is that at a given feed flow rate and feed gas temperature, the axial and radial spreading of the feed gas decreases with the increase of wall pressure, and at 100 m-bar wall pressure the axial spreading continues up to the $0.141\ Z$, with radial spreading up to $0.733\ R_{wall}$ ((Pradhan & Kumaran, *J. Fluid Mech.*, vol. 686, 2011, pp. 109-159); (Kumaran & Pradhan, *J. Fluid Mech.*, vol. 753, 2014, pp. 307-359)). Here, Z is the effective counter-current length, $R_{wall}$ is the radius of the cylinder, $V_{\theta,\ solid\ body\ rotation}$ is the angular velocity corresponding to solid body rotation, and $V_\theta$ is the actual angular velocity of the rotating gas after feed injection at a given axial and radial location. The DSMC simulation result indicates that the radial velocity is symmetric around the center-line of the feed point, and decreases along the radial and axial directions following a polynomial function. The DSMC simulation result also reveals that with the increase of axial feed location away from the top baffle, the radial profile of the axial mass flux becomes more flatten at the axial mid plane, and effects the counter-current circulation rate to a great extent.



**Keywords**: High-speed rotating flow, DSMC Simulations, Axial and radial spreading of the feed gas.


**References:**


[1] PRADHAN, S. & KUMARAN, V. 2011 The generalized Onsager model for the secondary flow in a high-speed rotating cylinder. *J. Fluid Mech.* 686, 109.

[2] KUMARAN, V & PRADHAN, S. 2014 The generalized Onsager model for a binary gas mixture. *J. Fluid Mech.* 753, 307.

[3] OLANDER, D. R. 1981 The theory of uranium enrichment by the gas centrifuge. *Progress in Nuclear Energy.* 8, 1.

[4] WOOD, H. G. & SANDERS, G. 1983 Rotating compressible flows with internal sources and sinks. *J. Fluid Mech.* 127, 299.

[5] WOOD, H. G. & BABARSKY, R. J. 1992 Analysis of a rapidly rotating gas in a pie-shaped cylinder. J. Fluid Mech. 239, 249.

[6] WOOD, H.G. & MORTON, J.B. 1980 Onsager's pancake approximation for the fluid dynamics of a gas centrifuge. *J. Fluid Mech.* 101, 1.

[7] BIRD, G. A. 1994 Molecular gas dynamics and the direct simulation of gas flows. *Clarendon Press*, Oxford.

[8] BIRD, G. A. 1963 Approach to translational equilibrium in a rigid sphere gas. Physics of fluids 6, 1518.

[9] BIRD, G. A. 1981 Monte-Carlo simulation in an engineering context. In S. S. Fischer, editor, Rarefied Gas Dynamics, proceedings of the 12th international symposium - part-1., 239.

[10] BIRD, G. A. 1987 Direct simulation of high-vorticity gas flows. Phys. Fluids. 30, 364.

[11] CHAPMAN, S. & COWLING, T. G. 1970 The Mathematical Theory of Non-Uniform Gases, 2nd edition. Cambridge, Cambridge University Press.

[12] CERCIGNANI, C. 2000 Rarefied gas dynamics. From basic concepts to actual calculations. Cambridge University Press.

## 231a Integrated Conceptual Design for Helium Recovery from Natural Gas

*Ying Liu, Qiang Xu*

Helium is an increasingly important inert gas in the modern economy, with a diverse range of applications across various industries. As natural gas is the most abundant source of helium, it is imperative to study the helium recovery associated with natural gas processing in effective and efficient ways. In this paper, an integrated LNG and liquid helium production process (namely LBRL process) from natural gas feedstock has been developed, modeled, and examined. It includes four subsystems: (i) a liquefied natural gas (LNG) production plant (LPP); (ii) a boil-off gas management system (BOGM); (iii) a helium recovery production plant (HRPP); and (iv) a helium liquefication plant (HLPP). The complex employs natural gas as the primary feedstock. Through process integration and optimization, the complex will produce LNG, liquid helium, and pure N2 simultaneously. The developed LBRL complex has been virtually demonstrated by rigorous modeling and simulations. Its energy consumption and economic performance under large turndown ratios have also been thoroughly investigated.

## 231b *Chemical Sector Risk Management Agency Resources for Enhanced Security and Resilience of the Chemical and Petrochemical Industries*

*Ashley Pennington, Janine Mason, Cheryl Louck, Jacob Mehl*

*The US Department of Homeland Security's Cybersecurity and Infrastructure Security Agency (CISA) functions as the Sector Risk Management Agency (SRMA) for the chemical sector. The Chemical sector is one of the 16 critical infrastructure sectors[(1)] whose assets, systems, and networks, whether physical or virtual, are considered so vital to the United States that their incapacitation or destruction would have a debilitating effect on security, national economic security, national public health or safety, or any combination thereof. Across CISA, teams work to support functions of the Chemical SRMA to ensure safe, secure, and resilient functioning of the chemical and petrochemical industries during times of serenity and during hazards or threats. The team will present an overview of the Chemical SRMA and its role in capacity building, training, publications, voluntary programs, and no-cost resources available to the chemical sector and within the four industry segments (basic chemicals, specialty chemicals, agricultural chemicals, and consumer products). The Chemical SRMA focuses on coordination and collaboration across the sector and with cross-sector critical infrastructure partners, both at the headquarters level and at the individual site level through the CISA regional offices. Among the resources of interest to the chemical and petrochemical industries discussed will be an overview of the 2022 Effective All-Hazards Chemical Sector Response Playbook[(2)] and webinar series which provides a Standard Operating Procedure (SOP) to assist the Chemical Sector in preparing for, responding to, and recovering from an all-hazards emergency. How to contact your regional office for interaction and resources, an overview of the Homeland Security Information Network – Critical Infrastructure (HSIN-CI) Chemical Portal information, attendance at classified and unclassified briefings, and how to receive pertinent Chemical SRMA updates will be covered.*

*(1) Critical Infrastructure Sectors | CISA*

*(2) Chemical Sector Playbook | CISA*

## 231c Batch to Continuous Transition of Large Volume Succinimide Dispersants

*Riddhesh Patel, Nasser Al Azri, Hari Mantripragada, Glenn Cormack, Nico Proust, Robert M. Enick, Goetz Veser*

The specialty chemicals industry to-date is largely operating using decades old technology based on large-volume batch reactors of tens to hundreds of thousands of gallons in capacity. This technology is simple but inefficient given the advances in science and engineering over the years. There is a large opportunity to improve the cost, energy, and atom efficiency of the processes and reduce their environmental footprint. Hence, the industry is currently undergoing a paradigm shift, moving away from traditional manufacturing approaches to intensified, modular, and cleaner processes.

Succinimide dispersants are widely used in the automotive industry—where they form a key component of engine oils—and in the production of emulsifiers for a wide range of applications, including explosives and metalworking. These highly viscous dispersants are formed via a two-step reaction between Poly-isobutylene succinic anhydride (PIBSA) and polymeric amines. They are traditionally produced in large-volume batch reactors (10-50k gal) with long batch times (~6 hours). Previously, we successfully demonstrated the transition to continuous production of succinimide dispersants using an integrated reactor-separator system. However, to-date, the production of the process feeds is still carried out in batch reactors which prevents the overall process from reaping the full benefits of continuous process operation.

The present study is investigating the production of PIBSA from polyisobutylene (PIB) and maleic acid anhydride (MAA). This reaction poses several challenges for continuous processing: i) highly viscous and poorly miscible reactants; ii) slow kinetics at typical operating conditions which result in long reaction times (>10 hours) and hence require operation at high-temperature, high-pressure conditions (T>200C, P>250 PSIG); and iii) undesired side reactions due to the temperature-sensitivity of MAA and its high reactivity with acidic and basic functionalities. In the present work, we aim to address these issues via use of a suitably designed high-pressure, high-temperature continuous reactor system; improving the mixing via emulsification with the reaction product, PIBSA; and mitigating undesired solids formation to improve product yield. Key to these advances are kinetic investigations of PIBSA formation as guide for transitioning to continuous operation in a stirred tank reactor versus a tubular reactor. Overall, this work aims to back-integrate succinimide dispersants manufacturing for a continuous process operation.

# 231d Executable Digital Twin: Bringing Insight from Engineering Simulation to Operations- Physics Based Simulation, Automation and Machine Learning

*Ravindra Aglave, Peter Mas, Thomas Eppinger, Umberto Badiali*

In the energy and process industries, physics-based digital twins are gaining prominence in virtual equipment development. Using simulation in the digital twin, engineers can determine optimal equipment designs before a unit goes into production.

Until recently, digital twins could not be deployed at the operational level, as they could be too complex or slow to react to the speed of operation. This is changing quickly. In this webinar, you'll learn how physics-based simulation lowers the barrier for executing the digital twins on the production floor.

**Using simulation to understand complexity**

Energy and chemical industry processes are complex. Using simulation to understand complexity creates opportunities to deploy the insights to your businesses' advantage. Physics-based simulation models, at varied length scales, allow engineers to capture this complexity and use it to explore design and operating spaces to find optimum conditions.

**Deploying insights from simulation into operations**

Deploy simulation insights into operations using executable digital twins. With digital twins, your engineers can determine the ideal design before the unit goes into production. Operations engineers can react to real-time product disturbances based on the deep insights gained during R&D.

**Key requirements for physics-based simulation**

The need to investigate problems across all lengths and time scales is one of the essential requirements for physics-based simulation. An engineering team may need a fit-for-purpose tool for a given scale. If multiple tools are required, integration must be seamless while offering a simple-to-use workflow.

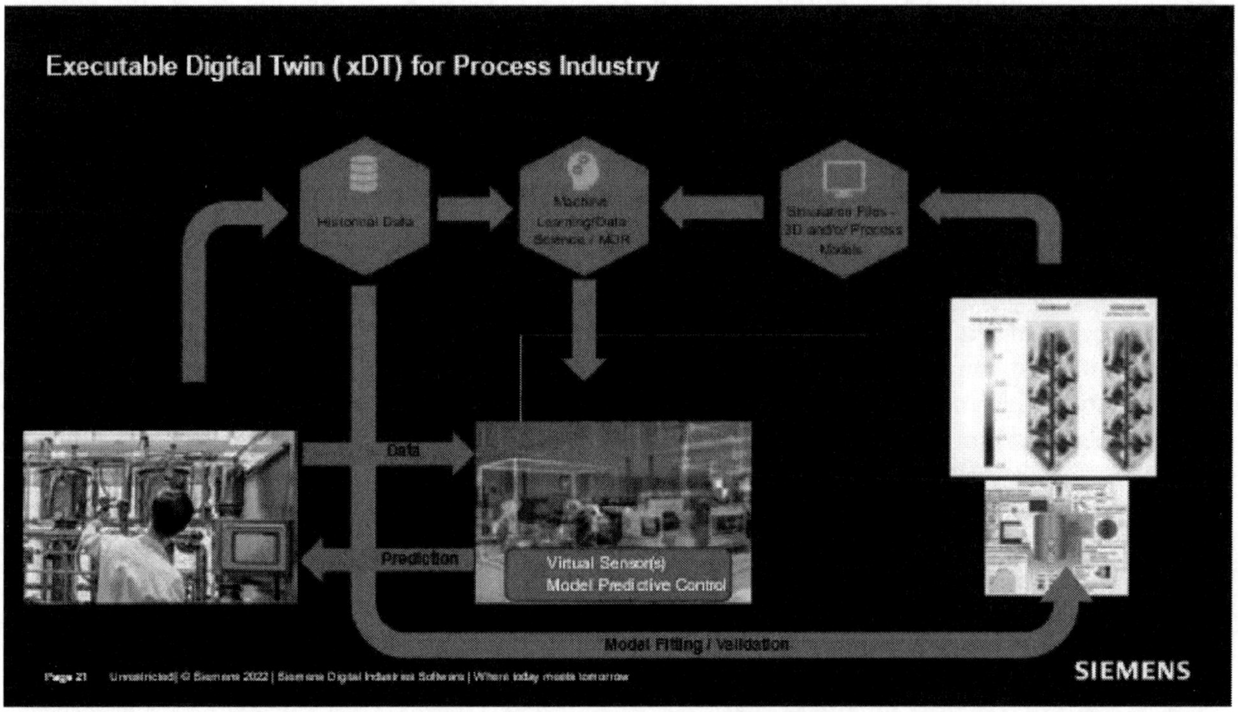

## 231e Optimal Design and Operation of an Ngcc Plant Integrated with Carbon Capture, Hydrogen Production and Long-Term Storage

*Zhien Zhang, Stephen Summit, Nishant Vinayak Giridhar, Md Emdadul Haque, Debangsu Bhattacharyya*

With an increase in the penetration of renewable energy into the electric grid, variability of load of fossil fuel-based power plants is expected to keep rising. Additionally, it will be desired to decrease the amount of $CO_2$ emitted by fossil fuel-based power plants through post combustion capture (PCC) systems. For accommodating high penetration of renewables into the grid, fossil fuel-based power plants will have to rapidly ramp up and down their loads with frequent shutdown. Such extreme load following conditions can cause higher wear and tear of critical components in the power plant, lower efficiency, and higher $CO_2$ emission due to decreased efficiency. Furthermore, renewables are intermittent and, in particular, solar energy may be limited for several months in some regions. For reducing extreme load-following operation of power plants, improve their efficiency and for facilitating long term energy storage, one potential option is hydrogen generation and storage. Hydrogen can be produced by using electrolyzers when the electricity is in abundance while utilizing it when there is high energy demand. For retrofitting with the existing natural gas combined cycle (NGCC) plants, $H_2$ can be co-fired with natural gas thus generating lesser amount of $CO_2$ and leading to lower penalty for $CO_2$ capture and compression, which facilitates generation and more power when the electricity price is high thus improving the economics.

In this work, a dynamic model of the proposed configuration is developed by including a model of the NGCC plant that can simulate varying extent of $H_2$ co-firing with natural gas, a model of the solvent-based PCC that facilitates flexible capture, and a model of the $H_2$ generation process by using a proton exchange membrane (PEM) electrolyzer stack, and a model of the $H_2$ compression and storage system. For $H_2$ storage, above ground pressure vessels and caverns are considered as alternative options. An economic model of the integrated process is developed by including capital costs for the PCC, PEM electrolyzer and the hydrogen storage and fixed and variable operating costs of these units. Carbon dioxide emissions are penalized using a carbon tax. We undertake net present value (NPV) optimization in Python/PYOMO for 14 market regions that differ greatly in terms of locational marginal price (LMP) of electricity and carbon tax rates. Hourly data of LMP for a duration of one year is considered for NPV optimization. Decision variables include design variables such as the capture plant capacity, dimensions of storage systems, electrolyzer capacity, etc. as well as hourly operating profile of the power plant, capture plant, and the hydrogen system considering charging, discharging and idling as possible options. The underlying NPV optimization problem is a large-scale, highly nonlinear dynamic optimization problem. For computational tractability, several strategies are developed including model order reduction, surrogate model development, model reformulation, etc. The optimization algorithm also allows the plant to shutdown if needed when certain constraints are satisfied.

We compare the economic performance of multiple flowsheet configurations including NGCC only, NGCC+PCC, NGCC+PCC+$H_2$ generation and storage. We identify candidate regions based on forecasted LMP trends, where the integrated energy configurations will be effective. In regions where cost reductions are necessary, we identify the minimum reductions in operating and capital costs for the integrated system to be economically viable.

**References:**


1. Yan, Y. *et al.* Harnessing the power of machine learning for carbon capture, utilisation, and storage (CCUS) – a state-of-the-art review. *Energy Environ. Sci.* **14**, 6122–6157 (2021).
2. Wang, Y., Bhattacharyya, D. & Turton, R. Evaluation of Novel Configurations of Natural Gas Combined Cycle (NGCC) Power Plants for Load-Following Operation using Dynamic Modeling and Optimization. *Energy and Fuels* **34**, 1053–1070 (2020).
3. Schill, W. P., Pahle, M. & Gambardella, C. Start-up costs of thermal power plants in markets with increasing shares of variable renewable generation. *Energy* **2**, (2017).
4. Cheng, F., Patankar, N., Chakrabarti, S. & Jenkins, J. D. Modeling the operational flexibility of natural gas combined cycle power plants coupled with flexible carbon capture and storage via solvent storage and flexible regeneration. *J. Greenh. Gas Con.* **118**, 103686 (2022).
5. Zantye, M. S., Arora, A. & Hasan, M. M. F. Renewable-integrated flexible carbon capture: A synergistic path forward to clean energy future. *Energy Environ. Sci.* **14**, 3986–4008 (2021).
6. Li, C., Conejo, A. J., Siirola, J. D. & Grossmann, I. E. On representative day selection for capacity expansion planning of power systems under extreme operating conditions. *J. Electr. Power Energy Syst.* **137**, 107697 (2022).
7. Scott, I. J., Carvalho, P. M. S., Botterud, A. & Silva, C. A. Clustering representative days for power systems generation expansion planning: Capturing the effects of variable renewables and energy storage. *Energy* **253**, 113603 (2019).
8. Bristowe, G. & Smallbone, A. The key techno-economic and manufacturing drivers for reducing the cost of power-to-gas and a hydrogen-enabled energy system. *Hydrogen* **2**, 273–300 (2021).

# 231f *Removal of Trace Amounts of Polluting Metals from Concentrated Electrolytic Rock Salt Solutions By Coagulation, Sedimentation and Ion Exchange*

*Benjámin Csorba, János Madarász, László Farkas, Renáta Zsanett Boros, Andrea Mihalkó, Marcell Csécsi, Iván László Gresits*

The previously widespread mercury cell technology for chlorine production has now been replaced by a more environmentally friendly membrane cell electrolysis, which is considered as BAT technology. However, this technology requires a rock salt solution that is much cleaner than before and contains polluting metals (Al, Ca, Mg, etc.) in the order of ppb (parts per billion, expressed in mass) at most. However, mined rock salt and sea salt contain a significant amount of these metal contaminants, so it is a very important task nowadays to remove the polluting metal content of the concentrated salt solution produced as a raw material of chlor-alkali industry as completely as possible. This is partially solved with current technology, but more efficient cleaning procedures promise many advantages. Thus, for example, they can increase the lifetime of the membrane of the electrolyzing cell, reduce the specific electricity consumption and contribute to the production of cleaner products, especially cleaner sodium hydroxide. Among the polluting metal ions aluminum is currently the most significant problem.

We previously developed a measurement method for the photometric determination of the trace amount of dissolved aluminum in concentrated salt solutions. The total (partly ionic, partly colloidal) aluminum content as well as other metal content can be measured using the ICP-OES method. Using these two methods, the sedimentation of metal contaminants in the untreated rock salt solution, as well as the coagulation in the case of aluminum, examining the time and pH dependence of the process can be followed. We determined the favorable pH values for the removal of individual metal ions, as well as the necessary sedimentation time at the operating temperature of 60-65 °C. In the case of aluminum, the surprising conclusion is that in addition to the neutral and weakly acidic pH around 5-7, the alkaline pH of 11 is also very favorable for the process in the case of the tested rock salt solution, compared to the case of laboratory-grade rock salt and the dilute aqueous medium. Therefore, we also examined the effect of other metal contaminants present. Based on our experiments, while iron and silicon impurities had no effect, in the presence of at least 150 ppm (parts per million, expressed in mass calcium or 5 ppm magnesium, the aluminum content measured at the top of the solution showed a similar decrease as in the case of untreated rock salt. The coagulation and sedimentation of aluminum was even stronger in the presence of a higher concentration or combined presence of these alkaline earth metals in appropriate concentrations.

The precipitates separated from the model solutions containing various metal impurities, as well as the real plant precipitates of BorsodChem Ltd. (the leading company of the chlor-alkali industry in Hungary), were also subjected to XRD tests. These tests showed that hydrocalumite is separated from a solution with a suitable calcium content, and hydrotalcite is separated from a solution with a suitable magnesium content, which are mixed salts of aluminum with calcium and magnesium (the anions are hydroxide and chloride). These tests also confirmed that, in the case of pH=11, significant precipitation of the aluminum content is caused by the presence of calcium and magnesium ions, through co-precipitation in a common compound.

In addition, the ability of aluminum-containing precipitates separated from different media to redissolve in acidic (pH=3) and alkaline (pH=11) salt solutions was also examined. Based on our results, the calcium and magnesium content of the original brine, and thus the resulting precipitate has a significant effect on the solubility of the precipitates in both acidic and alkaline salt solutions, compared to the solubility of the precipitate separated from a salt solution containing no calcium and magnesium.

So, the dissolved calcium and magnesium content has a beneficial effect on the removal of the aluminum contaminant in the brine from several points of view. On the one hand, it promotes the coagulation and sedimentation of aluminum by forming a common precipitate, and on the other hand, it also reduces the solubility of the precipitate, even in acidic saline solutions. Our results are confirmed both by the concentration of the solutions measured photometrically and by the ICP-OES methods, and by the XRD results of the precipitates. According to our experiments, individual contaminating metal ions facilitate the removal of each other, so compared to the current process, however only a pH change is sufficient to significantly improve the process.

The removal of aluminum ions remaining in solution after sedimentation by ion exchange was also investigated. Therefore, we conducted experiments with resin types containing different functional groups, mapping the dependence of the aluminum binding capacity of these resins on the solution's pH, salinity and the presence of other ions. In addition, we also examined the time dependence of the binding, as well as the lowest concentration to which each type of resin can reduce the dissolved aluminum concentration of salt solutions at different saturation levels of the resin. In addition, we also performed dynamic experiments mapping the effect of flow rates and the thickness of the resin bed, as well as investigating various regeneration methods. Our results contributed to the optimization of the operational ion exchange system.

Project no. C1340882 has been implemented with the support provided by the Ministry of Culture and Innovation of Hungary from the National Research, Development and Innovation Fund, financed under the KDP-2021 funding scheme.

NEMZETI KUTATÁSI, FEJLESZTÉSI
ÉS INNOVÁCIÓS HIVATAL

# 142a Removal of Red Tide Cells (*Karenia brevis*) and Toxins By Modified Biochar Flocculation

*Cadianne Chambers, Toufiq Reza*

With frequent Harmful Algal Blooms (HABs) events occurring destruction to marine ecosystems, and public health is of serious concern as microscopic algae lyze lethal toxins into freshwater and seawater. *Karenia brevis* also known as red tide has posed consistent threat to Florida waters such as Gulf of Mexico, and the southwestern coasts. Neurotoxin such as *Brevetoxin* are commonly produced from *Karenia brevis*, which causes fish kills, and distressing diarrhetic, neurologic, and respiratory symptoms. Thus, this upsurge adversely creates a negative impact on socioeconomic factors. This study navigates the use of biochar mixed with coagulant, polyaluminum chloride (PAC) to remove both the intra-cellular and extra-cellular (ruptured) brevetoxin cells. The modified biochar coagulants were synthesized via pyrolysis of at 500 °C for 1 h with varying biochar to PAC (0.5 mol/L) ratio of 1:5, 1:10 and 1:15. As biochar : PAC ratio is varied, observations on the material characterization were performed namely, ultimate and proximate analysis, thermogravimetric analysis, Brunauer-Emmett-Teller (BET) analysis, X-ray powder diffraction (XRD), point of zero charge ($pH_{PZC}$) and, Boehm titration analysis, and Fourier transformation infrared (FTIR) spectroscopy. Flocculation experiments were performed using slurry concentration of biochar coagulant (0.5 g/L) dispersed over known initial cell count of *Karenia brevis*. For this study, slurry concentrations were concocted by suspending biochar coagulant with deionized water. Ultimately, a 24 h study was performed to observe the behavior of *Karenia brevis* cells once interacted with the varying modified biochar coagulants. Analysis of remaining cells and ruptured toxins were identified by cell-counting and High-Performance Liquid Chromatography (HPLC) respectively. For the ruptured cells, microscopic examination were performed to confirm lysing of the cells. As a result, the investigated biochar coagulants showed versatility in mitigating Red Tide blooms ensuing adverse effects on the *Karenia brevis* cells and the dissolved brevetoxins.

## 142b Adsorption of Microcystin-LR on Model Hydrochar Studied Using Molecular Simulations

*Hemant Nagar, Toufiq Reza, Nirupam Aich, Sumit Sharma*

Microcystin-LR (MC-LR), the most well-known variant of the microcystin family, is a toxin that is produced by cyanobacteria during algal blooms. It is hepatoxic and can affect various human organs, such as the stomach, kidney, and brain, with the liver as a primary target. According to the World Health Organization, the maximum permissible limit of MC-LR in drinking water is 1 μg/L. Traditional water treatment processes such as chlorination and oxidization are costly for MC-LR treatment, and therefore, it is necessary to look cheaper ways of eliminating it from water supply. One solution is to use pyrolyzed hydrochar produced via hydrothermal carbonization for MC-LR adsorption from water. In this work, we have employed advanced molecular dynamics (MD) simulations in full atomistic detail to understand the adsorption behavior of MC-LR on hydrochar surfaces of different surface functional groups (ammonium, phosphate, and a charge neutral surface). Based on the surface charge, we note stark differences in the free energy of adsorption of MC-LR as well as the adsorbed orientation of the MC-LR molecule. We have identified key interactions that promote strong adsorption of MC-LR on hydrochar surfaces.

## 142c Exploring the Potential of Porous Iron Electrodes for Enhanced Electrocoagulation and Virus Removal: Insights from Neutron Computed Tomography and Energy Analysis

*Gyoung Gug Jang, Kyungho Kim, Shankar Chellam, Costas Tsouris*

In this study, we are interested in porous electrodes due to their potential to improve the efficiency of electrocoagulation (EC) processes. Possible improvements include reduced electrical energy consumption, enhanced mass transfer and pollutant precipitation, and effective virus removal. The aim of the study is to investigate the performance of porous iron electrodes in EC systems using advanced characterization techniques and energy analysis. Neutron computed tomography is employed to visualize and understand the electrode dissolution process, providing valuable insights into the dynamics of the porous structure during operation. Specifically, image analysis reveals the thickness and volume of each electrode, indicating whether the inner surface is effectively involved in EC reactions. Moreover, an in-depth material and energy analysis is conducted to quantify the energy consumption of the EC system with porous iron electrodes, offering a comprehensive understanding of energy efficiency improvements compared to conventional non-porous electrodes. Additionally, this study evaluates the performance of the EC system in removing viruses from water, focusing on the role of porous iron electrodes in enhancing virus removal efficiency. The findings of this research not only contribute to our understanding of the benefits of using porous iron electrodes in electrocoagulation systems but also highlight the potential of this technology for sustainable and effective water treatment applications.

## 142d Removing Pfas from Contaminated Water Using Apple Pomace Derived Cationic Cellulose Nanocrystal Coated Fibrous Filters

*Luis Franco Jr., Thomas Stuart, Md Shahadat Hossain, Bandaru V. Ramarao, Deepak Kumar*

Per- and poly-fluoroalkyl substances (PFAS) is a diverse chemical class with broad and extensive industrial applications. High aqueous solubility and low biodegradability have led to widespread PFAS water contamination, presenting serious health and safety concerns. PFAS contaminated water is commonly treated by granular activated carbon (GAC) or anion exchange resins. However, conventional adsorbents have limited efficacy against emerging short-chain PFAS species or in the presence of dissolved organic matter. This project aims to develop innovative high-capacity filter media for the treatment of PFAS-contaminated water using cellulose nanocrystals (CNC) derived from apple pomace, an apple processing waste product. CNC will be extracted from apple pomace, through pretreatment and acid hydrolysis. To enhance negatively charged PFAS retention, cationic protein aqueous extract (CPAE) derived from *Moringa oleifera* seeds will be coated onto CNC (CCNC). Batch adsorption studies will be performed for the removal of short-chain (heptafluorobutyric acid - PFBA) and long-chain (perfluorooctanoic acid – PFOA) PFAS molecules. Short- and long-chain PFAS adsorption rate, capacity, and robustness against solution conditions will be evaluated. For practical purposes, CCNC will be coated onto filter paper and tested by challenge PFAS solutions to quantify the removal efficiency for novel, biodegradable PFAS filter.

## 142e Selective Removal of Selenium from High-Salinity Wastewater Using Metal-Organic Frameworks

*Sweta Modak, Milad Esfahani*

Selenium shows the tendency of bioaccumulation in the food chain and shortens the life span of aquatic and wild animals. The concentration of selenium greater than 0.05 ppm is hazardous to human health; hence, removing it from drinking water is crucial. In addition, selenium is one of the components of the produced water (the waste stream in the oil extraction process with high salinity (ranging from 5000 ppm to 300 ppt)), which limits the reuse of the treated stream. Metal-organic frameworks (MOFs) are gaining significant attention recently as adsorbents due to their high porosity, tunability, and selectivity. In this study, the nickel-based ionic MOF (Ni-iMOF), zirconium-based MOF (MOF-808), iron MOF(Fe-MOF), copper MOF (Cu-MOF), and graphene oxide immobilized MOFs (GO-Fe-MOF and GO-Cu-MOF) were synthesized via different methods (such as solvothermal and acid-free-nonthermal process) using several metal salts and ligands and examined for the removal of selenium from water and saline water. The isotherm adsorption tests were conducted for all the MOFs (500 ppm) with 60 ppm selenium. The Ni-iMOF showed 15.48% selenite and 26.36% selenate removal, MOF-808 showed 62.44% selenite and 37.54% selenate removal, Fe-MOF showed 44.14% selenite and 1.53% selenate removal, GO-Fe-MOF showed 40.76% selenite and 1.64% selenate removal, Cu-MOF showed 1.18% selenite and 1.02% selenate removal and lastly, GO-Cu-MOF showed 0.82% selenite and 0.78% selenate removal. Among all the MOFs, the MOF-808 and Fe-MOF showed better adsorption of selenite and selenate in water. The better performance of MOF-808 is attributed to the coordination bonding between the charged metal in the MOF nodes and selenite and selenate molecules. To envisage the selectivity of MOFs toward selenium and selenate, isotherm adsorption tests were conducted in the presence of different salts ($Na_2SO_4$ (2000 ppm) and $NaHSO_3$ (2000 ppm)) for MOF-808 and Fe-MOF. The data shows that MOF-808 could remove 53.55% of selenite and 2.15% of selenate in the presence of $Na_2SO_4$ (2000 ppm), whereas it could remove 5.2% of selenite and 1.26% of selenate in the presence of $NaHSO_3$ (2000 ppm) salt in water. Similarly, Fe-MOF could remove 41.33% of selenite and 1.41% of selenate in the presence of $Na_2SO_4$ (2000 ppm) and it could remove 22.52% of selenite and 16.09% of selenate in the presence of $NaHSO_3$ (2000 ppm) from water. The effect of different experimental parameters such as pH (2-10), temperature (25-65°C), and initial adsorbate concentration ( 0.1-61 ppm) on the adsorption capacity of MOFs are currently under investigation.

# 142f High Recovery Desalination with a Flexible RO-NF Configuration Operating in the Mode of Partial Concentrate Recycle with Energy Recovery

*Yakubu Jarma, Beau Akhavain, Yoram Cohen*

Despite the maturity of reverse osmosis (RO) technology, management of the RO concentrate (i.e., "brine") and membrane fouling/mineral scaling remain as major impediments for inland deployment of RO desalination. High recovery RO desalting operation is critical, particularly in inland water desalination, for minimization of the volume of generated RO concentrate. However, high recovery RO requires a large number of RO membrane elements, which is a challenge particularly for the small, distributed desalination systems. Moreover, the concentration of sparingly soluble mineral salts, and suspended and dissolved foulant species (along the RO feed channel) with rising recovery which can exacerbate membrane mineral scaling/fouling. The consequence is reduced permeate productivity and increased operation and maintenance costs. In order to address the above challenges, a novel flexible system configuration that integrates nanofiltration (NF) and RO elements (Flex-RONF) was developed with the ability to operate in the mode of partial concentrate recycle, along with an installed pressure intensifier to reduce the required feed pump outlet pressure. Modeling of system operation and optimal membrane arrangement was assessed via process modeling, and the resulting configuration was evaluated for desalting of high salinity agricultural drainage (AD) water (~3,000-30,000 mg/L total dissolved solids) of high mineral scaling propensity (primarily gypsum and calcite). The study demonstrated that optimal configuration of NF and RO elements in a given system, along with a pressure intensifier, can effectively reduce the required applied pressure suitable for achieving an acceptable level of permeate quality, while also significantly reducing the system footprint relative to conventional RO system design. In order to reduce mineral scaling, automated periodic freshwater (permeate) flush was implemented at an optimal frequency. Evaluation of the FLEX-RONF system and the implemented operational protocol suggest the applicability of the approach for distributed water desalination given the reduced system footprint and high recovery operation.

# 142g Process Design of a LiBr Absorption Chiller Integrated Hdh-Mee-Mvr System for Zero Liquid Discharge Seawater Desalination

*Hasan Nikkhah, Burcu Beykal*

The world is facing a major challenge in providing access to clean water, with almost 70% of the population expected to face water scarcity by 2025 [1]. Desalination is a solution for producing potable water on a global scale [2,3], and among various thermal desalination techniques, humidification-dehumidification (HDH) is particularly suitable for small-scale desalination applications. However, to adapt this technology for large-scale production, the integration of HDH with other desalination technologies such as multi-effect evaporation (MEE) is necessary [3]. While this can increase the processing capacity, it can also increase energy consumption and $CO_2$ emissions. Several studies have also aimed to reduce the energy consumption of seawater desalination by integrating heat pumps, mechanical vapor recompression (MVR), and lithium bromide (LiBr) chillers [4,5]. Other studies have focused on reducing brine discharge or achieving zero liquid discharge (ZLD) [6]. However, there is a lack of a holistic study that simultaneously addresses these two critical environmental issues.

In this work, we design a multipurpose seawater desalination process that integrates a LiBr absorption chiller, HDH, MEE, and MVR to improve the energy effectiveness of conventional desalination processes [7]. This process not only decreases energy consumption but also aims to turn the desalination into a multipurpose unit by providing air conditioning, reducing the brine discharge, and preparing it for the ZLD process. The LiBr absorption chiller cycle preheats the seawater prior to the steam generation in the MEEs, while the MVR system activates the LiBr cycle. The integrated HDH system further recovers the excess water in the process stream and concentrates the brine effluent for ZLD processing. We comparatively study a stand-alone HDH desalination with our integrated system, quantify their performances over several metrics, perform a technoeconomic analysis to assess the economic feasibility, and carry out a sensitivity analysis to observe the effects of changing process variables on the overall performance of the integrated system. The results show that our proposed process design achieves several times higher gain-output ratio compared to a conventional HDH system of equal capacity, with a levelized water production cost of $8.4/m$^3$ and 6 kW of cooling capacity that can be used for air conditioning.

**References**

[1] Anand, B., Shankar, R., Murugavelh, S., Rivera, W., Prasad, K.M. and Nagarajan, R., 2021. A review on solar photovoltaic thermal integrated desalination technologies. Renewable and Sustainable Energy Reviews, 141, p.110787.

[2] Panagopoulos, A., 2021. Water-energy nexus: desalination technologies and renewable energy sources. Environmental Science and Pollution Research, 28(17), pp.21009-21022.

[3] Lawal, D.U., Antar, M.A. and Khalifa, A.E., 2021. Integration of a MSF desalination system with a HDH system for brine recovery. Sustainability, 13(6), p.3506.

[4] Lawal, D.U., Antar, M.A., Khalifa, A., Zubair, S.M. and Al-Sulaiman, F., 2020. Experimental investigation of heat pump driven humidification-dehumidification desalination system for water desalination and space conditioning. Desalination, 475, p.114199.

[5] Xu, H., Zhao, Y., Jia, T. and Dai, Y.J., 2018. Experimental investigation on a solar assisted heat pump desalination system with humidification-dehumidification. Desalination, 437, pp.89-99.

[6] Tahir, F. and Al-Ghamdi, S.G., 2022. Integrated MED and HDH desalination systems for an energy-efficient zero liquid discharge (ZLD) system. Energy Reports, 8, pp.29-34.

[7] Nikkhah, H. and Beykal, B., 2023. Process Design and Technoeconomic Analysis for Zero Liquid Discharge Desalination via LiBr Absorption Chiller Integrated HDH-MEE-MVR System. Desalination (*Under Review*).

# 202a Utilization of Nuscale Voygr NPP Integrated Energy System (IES) for the Production of Decarbonized Ammonia

*Amy Kozel, Francis Tsang, Jose Reyes, Luis DePavia*

Ammonia is a highly versatile chemical widely used for industrial purposes and prevalent in our everyday lives. The industrial applications of ammonia are vast and diverse. One of its primary uses is as a key ingredient in fertilizers. Ammonia is also used in the production of numerous chemicals, including plastics, textiles, and pharmaceuticals. Its ability to bond with water makes it an ideal refrigerant, used in large-scale cooling systems such as those found in food storage and processing facilities. Its unique properties and diverse range of applications make ammonia a critical component of modern industry.

Ammonia is almost exclusively produced using the Haber-Bosch process, which reacts nitrogen ($N_2$) and hydrogen ($H_2$) with a heterogeneous metal catalyst to produce ammonia ($NH_3$). Currently, the predominant method of producing the necessary hydrogen is the steam-methane reforming (SMR) process in which natural gas containing methane ($CH_4$) is reacted with steam. Unfortunately, this process results in significant $CO_2$ emissions into the atmosphere.

A decarbonized approach to generate $NH_3$ would be to use an Integrated Energy System (IES) that includes a NuScale VOYGR nuclear power plant (NPP) for electric power and steam production. The IES would be coupled to a High-Temperature-Steam-Electrolysis (HTSE) system containing Solid Oxide Electrolysis Cell (SOEC) stacks for hydrogen and oxygen production. The following is a listing of the functions of the subsystems and the intermediate materials for the entire production process as well as the expected NPP inputs:

- Reverse Osmosis (RO) Water Treatment – Clean water into steam generator to produce steam (power)
- Steam Generator – High quality steam into HTSE for hydrogen production (steam)
- HTSE System – Hydrogen and oxygen production (power)
- Nitrogen Generation – Air separation (power)
- Ammonia Production – Reacting hydrogen and nitrogen to produce ammonia (power)

**RO Water Treatment –** NuScale internal research has demonstrated that RO technology when coupled with NuScale VOYGR NPP generate the smallest carbon footprints (g-$CO_2$ per $m^3$ of clean water) when compare to other energy sources including renewable and hydropower.

**Steam Generation –** NuScale has studied the control strategies needed to divert a portion of the steam produced (in its secondary process, isolated from radiation) away from the steam turbine for power production and towards providing thermal energy to an adjacent process. In this application, the superheated steam is used to generate saturated steam to feed to the HTSE system.

**Hydrogen Production -** NuScale has teamed up with DOE National Lab and technology companies to investigate in using NuScale VOYGR NPP for the production of Hydrogen and Oxygen via the HTSE-SOEC technology (Ref 1 and 2). Other electrolyzer technologies can certainly be considered (such as Polymer Electrolyte Membrane, PEM, and Alkaline Electrolysis, AEL), but the HTSE system is particularly well suited in this application as the NPP provides both thermal and electrical energy, increasing its total system efficiency to 90% higher heating value (HHV). As indicated above, the thermal energy directly generates the steam feed to the HTSE and the electrical energy supplies power to the SOEC, feed gas compression to the Haber-Bosch process, and all balance of plant requirements.

**Nitrogen Production -** There are three methods for the generation of nitrogen from air components: a) Pressure-Swing-Absorption (PSA), b) membrane, and c) cryogenic systems. All the prescribed processes to generate nitrogen require significant amounts of energy (primarily electrical) to compress, heat/cool, and purify the $N_2$.

**Ammonia Production** – The Haber-Bosch process involves compressing the feed gases N2 and H2 to ~300 bar and ~450C. The feed gases enter a reactor containing a metal heterogeneous catalyst. The reaction is a mildly exothermic equilibrium reaction. The outlet gases are cooled to condense ammonia and the unreacted N2 and H2 recycled back to the reactor. Electrical power is used to compress the inlet gases and provide heating and cooling as needed.

NuScale Power is the industry-leading provider of proprietary and innovative advanced small modular reactor nuclear technology. It delivers scalable, and reliable carbon-free energy. NuScale VOYGR Small Modular Reactor Power Plant system's modularity and flexibility is uniquely qualified to support desalination operations, especially for large-scale production systems. NuScale VOYGR™ SMR plants are powered by the NuScale Power Module™—each a small, safe, pressurized water reactor that can generate 77 megawatts of electricity (MWe) or 250 megawatts thermal (gross), and can be scaled to meet operational needs through an array of flexible configurations up to 924 MWe (12 modules) of output.

The NuScale SMR is the only small modular reactor to have its design certified by the U.S. Nuclear Regulatory Commission and published in the Federal Register on February 3, 2023 (Ref 3). In addition, the standard Emergency Planning Zone (EPZ) for existing NPPs is a 10-mile radius. However, the NRC approved the NuScale (EPZ) sizing methodology, limiting the EPZ to the site boundary of the power plant for a wide range of potential plant sites where a NuScale VOYGR SMR power plant could be located. **This means that users of the plant's output can be located in close proximity.** Therefore, NuScale is well positioned to serve diverse customers across the world by supplying nuclear energy for electrical generation, district heating, desalination, commercial-scale hydrogen production, and other process heat applications.

The attached image shows Figure 1: the setup and the coupling of the NuScale VOYGR NPP to all the subsystems and components for the production of decarbonized ammonia and Figure 2: a detailed NuScale VOYGR plant site layout.

**References**


- Idaho National Lab Report, INL/LTD 14-31039, February 2014 (Limited Distribution)
- Idaho National Lab Report, INL/EXT-19-55395 Rev.1, September 2019
- https://www.federalregister.gov/documents/2023/01/19/2023-00729/nuscale-small-modular-reactor-design-certification

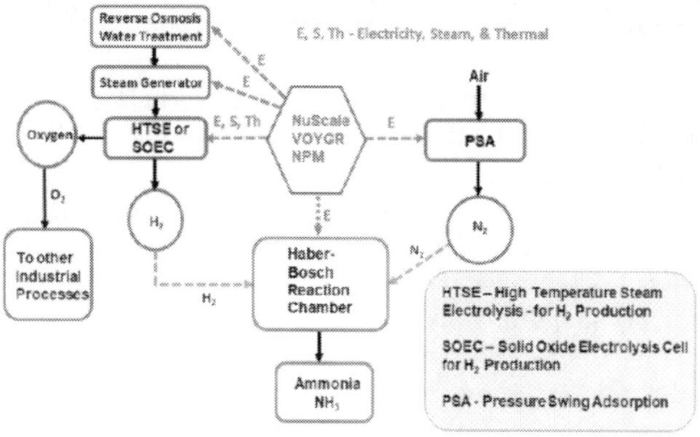


gure 1. NuScale VOYGR Plant NPP for Decarbonized Ammonia Production

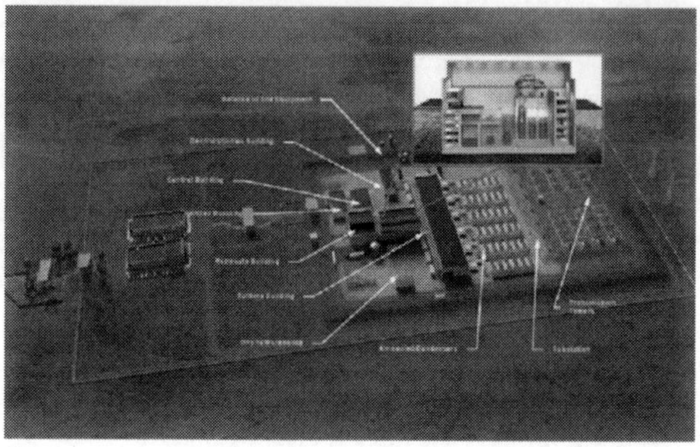

Figure 2. Detailed NuScale VOYGR Plant Site Layout

## 202b Renewable Ammonia for a Sustainable Energy Future: The Case of Maritime Transportation.

*Andrea Mio, Elena Barbera, Katerina Cufar, Alessandro Massi Pavan, Alberto Bertucco, Maurizio Fermeglia*

The energy sector is responsible for 75% of greenhouse gas emissions. We therefore want to reduce carbon dioxide emissions to reach net zero by 2050. The transition to an energy sector dominated by renewable sources also requires an appropriate growth of storage systems that make up for the difference between demand and production of electricity. The use of ammonia as an energy carrier can facilitate the energy transition. The high energy density, the ease of storage and transport, make it a very promising energy carrier. In addition, ammonia produced from renewable sources, i.e. water, air and renewable energy, could also be used for other applications: transport fuel, hydrogen carrier, long-term energy storage system, etc. Consequently, ammonia could play a crucial role in the decarbonization of several sectors, especially the naval sector. There are still several technological, economic and political challenges to be addressed for the production and use of ammonia produced from renewable sources.

In this work, two different ammonia production processes were simulated: the traditional one, i.e., the Haber-Bosh process using methane as raw material and a novel one based on green hydrogen production via electrolysis. Starting from the material and energy balances obtained by process simulation, CAPEX and OPEX were calculated and three indicators assessing sustainability performance have been evaluated. Well-established indicators, such as Energy Return On Energy Invested (EROEI), the Levelized Cost of Ammonia (LCOA), and the Life Cycle Assessment (LCA), allowed the comparison between the different processes simulated from an energetic, economic, and environmental viewpoint, respectively. The indicators' outcomes indicate that the most convenient process in terms of energetic performance is the green hydrogen-based with electrical energy obtained from a green energy mix, counterbalanced by a higher cost of ammonia produced via green hydrogen. Environmental assessment highlighted that a trade-off must be found in order to select the most environmentally friendly production pathway.

## 202c Investigation into Thermochemical Looping Process for Sustainable Ammonia Production

*Grant Bauer, Bharathkiran Maddipudi, Rajesh Shende*

Investigation into thermochemical looping process for sustainable ammonia production
Grant Bauer and Rajesh Shende*
Karen M. Swindler Department of Chemical and Biological Engineering
501 E. St. Joseph, Rapid City, South Dakota 57701
*Corresponding author email: Rajesh.Shende@sdsmt.edu' Rajesh.Shende@sdsmt.edu
Industrially, ammonia is produced by the Haber-Bosch (HB) process, which is energy and carbon intensive. In the HB process, nitrogen and hydrogen react over a supported Fe catalyst. It is to be noted
that hydrogen at a large scale is still generated by the steam methane reforming (SMR) process, which is
not sustainable as it generates carbon dioxide. In recent years, researchers are developing low energy and carbon intensity processes and investigating novel materials. A two-step chemical looping (CL) process involves nitridation of metal oxide (step-1) followed hydrolysis of metal nitride (step-2) to regenerate original metal oxide material. Typically, the metal oxide is regenerated at a lower temperature with the introduction of steam for the nitride hydrolysis producing ammonia.. We hypothesize that the use of binary or ternary mixture of transition metals (TM) oxide, can lower the nitridation temperatures. Novel binary and ternary transition metal oxides were designed and synthesized. A series of experiments are currently underway to understand efficacy of these materials towards ammonia production at lower temperatures. Results obtained with the synthesis, characterization and performance evaluation towards ammonia generation will be presented.

## 277a Durability of Anion Exchange Membrane $CO_2$ Electrolyzer When Exposed to Flue Gas

*Zengcai Liu, Hongzhou Yang, Michael G. Resch, David Cullen, Richard Masel*

Carbon dioxide ($CO_2$) is the largest contributor among the global warming gases and is responsible for increasing temperatures through a process called the greenhouse effect. One way to mitigate $CO_2$ is to convert $CO_2$ electrochemically into value-added chemicals such as C1 (carbon monoxide, formate/formic acid) and $C_2$+ products. Amongst these technologies, electrochemical reduction of $CO_2$ to CO is most mature technology in laboratory. Stable performance of $CO_2$ electrolysis to CO and $O_2$ for up to 4000hrs with negligible degradation has been demonstrated when operating with pure $CO_2$ [1,2]. However, industrial $CO_2$ sources usually contain other species such as nitrogen ($N_2$), Oxygen ($O_2$), nitrogen oxides ($NO_x$), volatile sulfur compounds (VSC: $SO_2$, $H_2S$, COS), volatile oxygenates ($VO_X$: ethanol) and volatile hydrocarbons (VHC). In this work, we investigate the effects of $H_2S$ up to 25ppm in $CO_2$ on the performance of anion exchange membrane $CO_2$ electrolyzers (AEMCEs). Figure 1 shows the stable performance of AEMCE for up to 3000hr. We will discuss the possible degradation mechanisms related to $H_2S$.

Figure 1 Long term stability of $CO_2$ electrolyzer running at 500mA (100mA/cm$^2$) in the presence of 3ppm of $H_2S$ in $CO_2$ stream

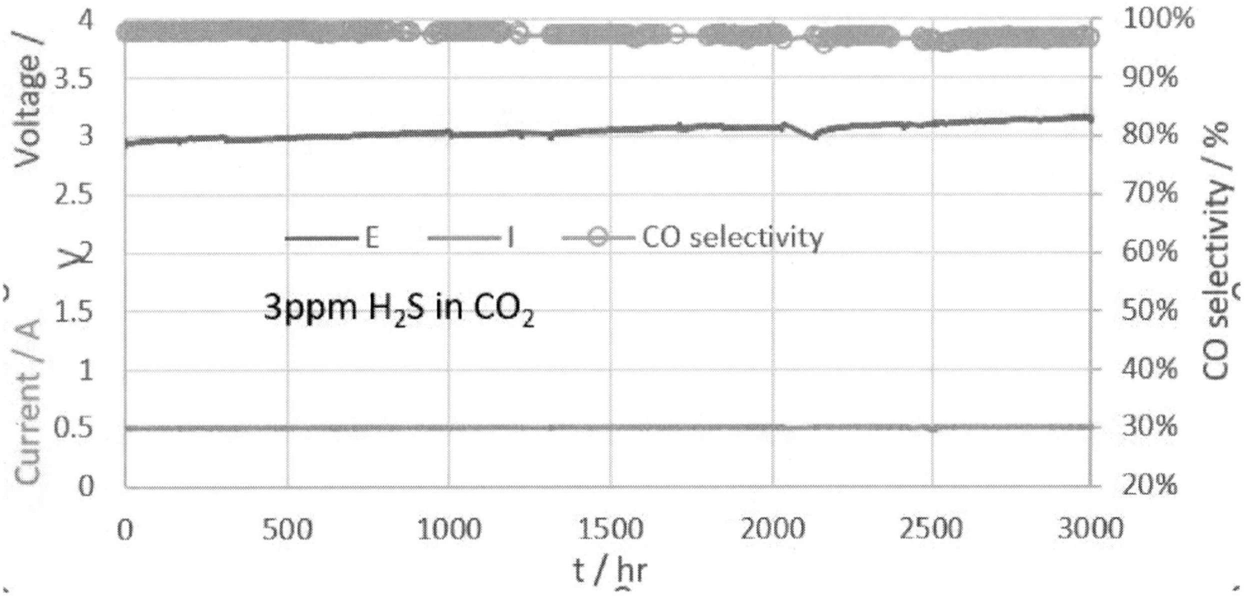

## 277b Comparative LCA for Decarbonization of Petrochemicals Using Electrification, Biomass or $CO_2$ As Feedstocks

*Elias Martinez-Hernandez, Juan C Ramos Consuelo, Diana Dominguillo Ramirez, Aburto Jorge*

Chemical industry decarbonization will require an in-depth analysis of direct GHG emissions (also known as Scope 1 emissions) to identify hot spots and the scope for mitigation measures. Furthermore, the chemical industry implies complex multiproduct value chains that will require systematic structural and performance evaluations for clear allocation of impacts in order develop strategies that also address scope 2 and 3 emissions (indirect emissions). To evaluate alternatives and strategies a life cycle approach needs to be adopted. In this work, the life cycle assessment (LCA) of alternatives for decarbonizing ethylene and methanol as major petrochemical products is presented. Ethylene and methanol are generally produced from fossil feedstocks using energy intensive and $CO_2$ emitting process technologies. In both cases the objective is to identify hot spots, impact trade-offs and reduction of life cycle GHG emissions (covering scope 1, 2 and 3 emissions). Because ethylene is the major petrochemical produced and thus a major contributor to greenhouse gas (GHG) emissions, its decarbonization is much needed. For this process, electrification of the crackers and the use of bionaphtha (from hydrodeoxygenation of vegetable oil) was considered. Compared to the use of fossil naphtha, the use of bionaphtha allowed a 15% decrease in GHG emissions reduction, while the electrification of the crackers using renewable energy allowed a 50% decrease. The reduction achieved by combining both decarbonization alternatives was 60%. However, trade-offs with impacts such as acidification and photochemical ozone creation were identified. A similar analysis was performed for the case of methanol and preliminary results will be presented. One alternative was the use of biomass gasification for syngas production and conversion into methanol. A second alternative included the use of $CO_2$ and green hydrogen as feedstocks. These alternatives are compared against the process using natural gas. Given the pressure on the chemical industry to set transparent targets, measure and monitor decarbonization, the approach and results presented can provide insights on the major bottlenecks and trade-offs of the decarbonization challenge from a whole system perspective.

## 277c High $CO_2$ conversion to CO Using Reverse Water Gas Shift at Ultra-High Temperatures

*Bradley Brennan*

Dimensional Energy has developed proprietary catalysts and reactor systems for high conversion of $CO_2$ + $H_2$ to CO + $H_2O$ using the Reverse Water Gas Shift (RWGS) reaction. The output gas mixture can be utilized as a source of *syngas* (CO + $H_2$) for downstream production of hydrocarbon products such alcohols or Fischer-Tropsch paraffins. This sustainable hydrocarbon formation is a Power-to-Liquids process where $CO_2$, water, and electricity are the only inputs. Utilizing ultra-high temperature ceramic reactors and optimized catalysts, $CO_2$ conversions were obtained of ~75%. Catalyst selectivities near 100% were obtained, and the syngas outputs were capable of being directly input into a Fischer-Tropsch reactor. The 100 kg per day $CO_2$ pilot experiment shows the scalability of the reactor system. Results will be presented for the chemical reactor and integrated pilot plant with Fischer Tropsch.

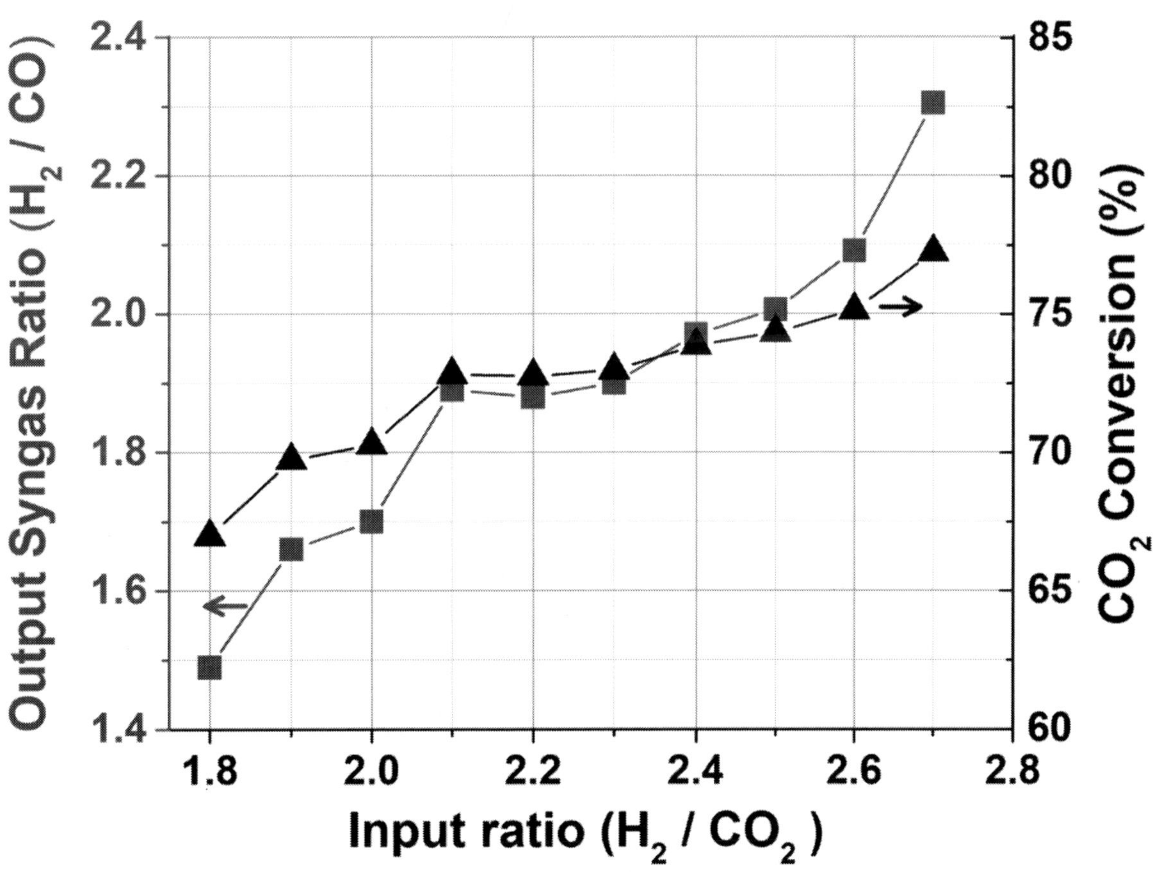

# 277d Multi-Dimensional Modeling of Inductively-Heatedsteam Methane Reforming Reactor

*Yufei Zhao, Cornelius Masuku, Chengtian Cui*

Steam Methane Reforming (SMR) accounts for 50% of the global $H_2$ production. Due to its endothermic property, heat source is required during production process. Current industrially used conventional SMR reactors employ a tube-furnace design. Extra methane (natural gas) is combusted for heat supply, leading to 1.7% of the total $CO_2$ emissions. To decarbonize the reactor, induction heating induced by alternating current has been investigated as a potential alternative. While lab-scale inductively heated reactors have been demonstrated, this study focuses on analytically and numerically characterizing the performance of such a reactor and evaluating its potential for industrial use. We propose a dynamic multi-dimensional heterogeneous mathematical model of an inductively heated SMR reactor. The 2D-1D model is based on the $1^{st}$ principle, with the reactor domain considering longitudinal and radial directions, while the catalyst domain considers radial direction. Electromagnetic heat including both eddy current and hysteresis heat from NiCo alloy acts as heat source instead of combustion. The model is solved using the finite volume method. Methane conversions under different flow rates are compared with literature lab-scale experimental results. The designed model is used for scale-up, and a detailed computational fluid dynamics simulation is performed. Finally, environmental analyses including power consumption, energy efficiency, and carbon reduction are quantified.

Overall, our study contributes to the decarbonization of conventional hydrogen production and provides a more sustainable and environmentally friendly energy system as reference for an endothermic reaction.

## 277e Optimizing Work and Heat Flows in Sustainable Chemical Processes Using Attainable Regions

*Diane Hildebrandt, Neil Stacey, James A. Fox, Baraka Celestin Sempuga, Maphala Ephraim Llane*

A chemical process aims to transform input materials into desired products. This transformation requires the supply or rejection of heat and work between the process and the environment. Careful selection of chemical pathways, process parameters, and heat/workflows is essential to optimize process performance. In pursuit of environmental sustainability, the upper limit for process performance is a reversible process, setting the targets for process design.

The GH diagram is a valuable tool for process synthesis and assessing process reversibility. This diagram plots enthalpy (H), which corresponds to the process heat flows, against Gibbs Free Energy (G), which corresponds to the process workflows. We can determine the heat and workflows required for any chemical process or pathway through the GH plot. The GH diagram offers the advantage of presenting and analyzing any number of processes or process steps with varying complexities on a single set of two-dimensional axes.

In order to represent all processes on the GH diagram, we first determine The *Material Balance Attainable Region* ($AR^{MB}$), which represents all possible product flow rates achievable for a given feed(s) for all possible processes. The dimension of the $AR^{MB}$ corresponds to the number of independent material balances describing the relationship between the feed and product species. However, the $AR^{MB}$ can be transformed into a two-dimensional convex projection in GH space, called the Thermodynamic Attainable Region ($AR^{T}$), which represents the set of heat and workflows for all possible processes converting the given feed(s) into any combination of defined products.

Leveraging the visual representation of the $AR^{T}$, we can systematically analyze and compare the inherent heat and workflows for different processes, feeds and products. To illustrate the impact of feed selection on heat and workflows, we will consider some simple examples. Additionally, we will compare the addition of work and heat using solar energy, either captured as heat or transformed into electricity. This analysis enables us to specify renewable energy mixes tailored to the chosen chemical pathway, minimizing cost and land use and facilitating quick comparison of the suitability of different processes for utilizing a known renewable energy mix.

## 168a Risk Mitigation in Process Scale up of Aminopropyl Silicone Intermediates

*Michael Telgenhoff, Brian Rekken, Xiaoyun (Shawn) Chen*

Within silicones, access to terminal aminopropyl silicone polymers has been limited due to the difficulties in synthetic methods of aminopropyl silicone endblocking intermediates. Bis-aminopropyl disiloxane is generally used to build terminal primary aminosiloxanes via a silicone equilibration process. However, this intermediate requires multiple synthetic steps and raw materials making production costly. In addition, the production chemistry requires based-catalyzed equilibration with dimethyl silicone followed by neutralization and filtration. An alternative synthetic method involves the primary amino, 2,2,4-trimethyl-1- aza- 2- silacyclopentane (CAS: 661492-94-0) or N-[(3-Aminopropyl)dimethylsilyl]-2,2-dimethyl-1-aza-2-silacyclopentane (CAS: 388606-32-4) as a direct reaction with silanol terminated silicones. However, synthetic yields have been low or a catalyst was required for the silanol capping reaction. Another route that was not previously investigated involves the use of cyclic-siloxazanes, specifically 2, 2, 7, 7- tetramethyl-1- oxa- 3- aza- 2, 7-disilacycloheptane. The cyclic siloxazane readily ring opens at the Si-N bond when combined with silanols to add a disiloxy-propylamine unit. This will occur at ambient temperature and free of catalyst. Prior to the work reported in this paper, the process to make the cyclic-siloxazane (GE Silicones) was plagued by low-volume efficiency and a hazardous, untenable process in a manufacturing setting. Attempts by GE to reduce excessive use of allylamine led to disastrous results —a runaway reaction. As it will be shown in this presentation, the large exotherm was likely a result of an accumulation of SiH and olefin functionalities in the mixture concurrent with the reduction of free allylamine, which is known to act as a hydrosilylation inhibitor. In this work, we show that the cyclic siloxazane can be successfully and safely synthesized through new process controlled methods. The scale up from a glove box to a 1L fume hood reactor with batch distillation was successfully demonstrated for the production of cyclic siloxazane endblocker. There are four key technologies specific to the hydrosilylation of olefinic amines in this example that are required and enable the sensitive chemistry. These include pulsed feed of tetramethyldisiloxane, use of a hydrosilylation promoter, ammonium sulfate catalyzed rearrangement, and batch distillation of the lights and product. Proper implementation of these methods can result in volume yields ~89% at >99% purity.

# 168b How Pilot Testing and Modularization Can De-Risk Technology Commercialization and Overall Project Delivery

*Chris Rentsch*

Developing a project execution strategy to commercialize developing technology projects requires a degree of sensitivity to its unique challenges. Project drivers that maximize the probability of advancing through process development phases and financial decision gates are vital in early development.

Unlike traditional projects, developing technology projects walk a delicate line between being economically viable or never making it past the conceptual stage. Having the right project execution strategy, which delivers cost & schedule certainty beginning early at the conceptual stage and maintains project certainty through process development and validation, engineering, construction, and facilities start-up, is fundamental to the success of developing technology projects. Two such best practices which drive project certainty are pilot testing and a modular project delivery model.

When commercializing a new technology, one best practice is to perform pilot testing to de-risk process and obtain a process performance guarantee prior to scale up and commercialization. As it relates to overall project execution, it is also important to select the best-suited delivery model to meet project goals and drivers. When it comes to executing developing technology projects, a truckable module project delivery model is well-suited to provide Owners and Developing Technology providers the project certainty needed to ensure the greatest probability of meeting project financing hurdles, providing cost and schedule advantages, and project certainty throughout the project lifecycle as compared to a traditional field construction project execution strategy. Furthermore, a modular delivery model provides Owners the flexibility to develop a modular pilot or demonstration-scale unit which can easily be commercially scaled and duplicated across multiple geographies.

Our paper will discuss these project delivery challenges inherent to developing technology projects, how pilot testing was planned and performed to de-risk the project, and when a truckable module delivery model is best suited as an execution model and its benefits compared to traditional field construction, and provide current relevant project examples and lessons learned. Project examples could include PureCycle Technologies, Origin Materials among others.

## Recycled Polypropylene Project Phase 1 – Demonstration Scale

**Client:** PureCycle Technologies
**Product:** Recycled Polypropylene
**Process:** Confidential

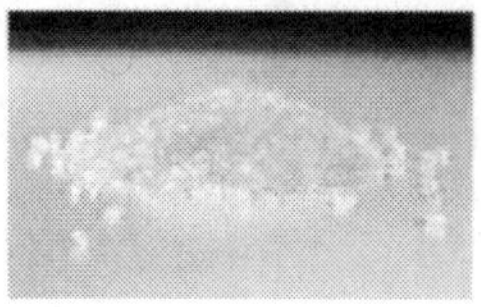

**Project Overview:**

Koch Modular worked together with PureCycle to development its advanced recycling technology which removes color, odor and other contaminants from recycled feedstock resulting in virgin-like ultra-pure recycled polypropylene (UPRP).

Koch Modular completed Phase I Feedstock Evaluation Unit and delivered successful production of UPRP at scale. Subsequently, Koch Modular has been awarded contract to construct Phase II of its commercial recycled polypropylene facility, to be located in Ironton, Ohio and designed to produce over 105 million pounds of UPRP per year.

Construction of the facility is underway, and the plant is expected to come online in 2022.

Koch Modular
PROCESS SYSTEMS 1

## Recycled Polypropylene Project Phase 2 – Commercial Scale

**Client:** PureCycle Technologies
**Product:** Recycled Polypropylene
**# Modules:** 25 (18 Process, 3 Support, 4 Field Erected) + 3 Stand-Alone Vessels & Pipe Racks.
**Schedule:** 18 months

*"Koch Modular's guidance, professionalism and end-to-end expertise in innovative technologies scale-ups like this one, plus their modular construction approach, was a critical success factor for our project."*

**MIKE OTWORTH**
PureCycle Technologies
CEO

Koch Modular
PROCESS SYSTEMS 2

## 168c Partial-Least-Squares Based Analysis to Manage Risks Identified for a Fluid Bed Wet Granulation Operation

*Maitraye Sen, Sydney Butikofer, Shashwat Gupta, Meng Zhao, Adam S. Butterbaugh*

An important step at the outset of any pharmaceutical process development is the process risk assessment that identifies the process variables carrying a higher risk towards the product quality attributes. Throughout the development phase, data is gathered to mitigate the risks. For this purpose, it is important to obtain a holistic understanding of the process while evaluating the importance of each variable (considered in the risk assessment) towards influencing the product quality.

This work presents a case study on a risk driven development of a fluid bed wet granulation operation followed by compression for making the tablets (drug product). The parallel mechanisms of fluidization, granulation and drying are coupled and dependent on one another that makes fluid bed wet granulation a complex process. In this study, multivariate perturbations via changing material properties and operating conditions were explored. A partial-least-squares model developed from the experimental data was used to evaluate the failure modes on the tablet critical quality attributes for more insight into what variable combinations could result in such situations. Finally, the conclusions from the partial-least-squares model were used to eliminate/manage some of the identified risks.

# 168d Digital Materials Engineering for Development of Materials

*Ravindra Aglave, Flavio Souza, Jan-Willem Handgraaf, Thomas Eppinger*

Large technological advancements are almost always dependent on the creation (or discovery) of novel materials. Revolutionary developments in material science enable the next generation of technological advancement. The full exploitation of those technologies (by engineers) will often require further investment into the evolutionary development of those materials. Advanced materials with novel properties hold the key to innovation for organizations across industries. As the transition to clean energy gathers pace, there's increased pressure on design engineers and researchers to develop innovative materials for sustainable energy solutions.

However, the path from initial discovery of new materials to production can be costly and fraught with many challenges. This paper examines how simulation can be an effective way for organizations to discover and develop new materials faster to solve their complex engineering challenges.

In this talk we will speak about the entire value chain of materials: from discovery to performance assessment to production processes as well as post-production processing. We will also identify the current gaps in this workflow as well as our activities to close those gaps. The Siemens solution consists of an end-to-end workflow based on the digital twin concept with four main steps: materials discovery, materials performance analysis, process design and engineering, and materials processing. The talk will include examples of these phases to weave a digital thread through the solution as well as the tools that are involved both for simulation as well as the backbone (material lifecycle management) platform for digital engineering.

Digital Materials Engineering

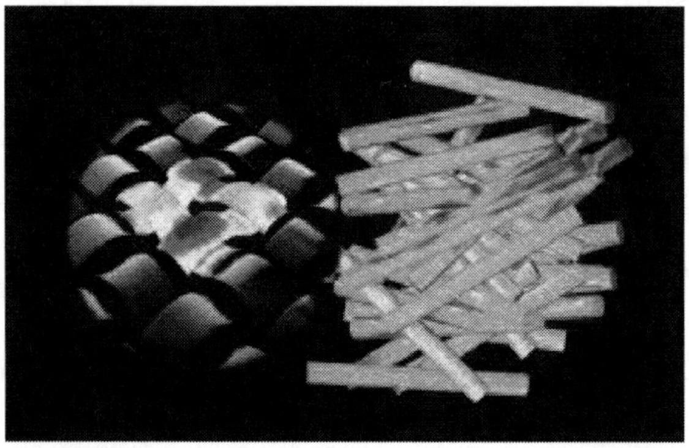

## 168e When to Stop Additional Attempts in a Particular Innovation

*Darrell Velegol*

The purpose of this article is to assess the Optimal Stopping point in innovation processes. How do we know when to stop trying new prototypes, new hypotheses, new experiments for a given innovation project? Optimal Stopping theory from probability analysis gives us some key results, and here I extend some of these to more realistic situations, finding several surprising results. It is proposed that organizations score their successive trials so that they can determine when to stop, so that they optimize their return on investment and reduce risk of loss. A practical Delphi method will be introduced for this scoring. This talk is an important example of an "algorithm for innovation".

# AUTHOR INDEX